Smart Systems

The focus of this book is to cover the fundamentals, methodological approaches, and diverse applications of smart systems. **Smart Systems** discusses important topics such as the Internet of Things—enabled smart systems, artificial intelligence, ergonomics, digital twin, and quality assurance frameworks in smart systems. It addresses methodological approaches in diverse sectors such as service, agriculture, product design, and development.

This book:

- Discusses important concepts such as customer satisfaction, product design, and product lifestyle management in a comprehensive manner.
- Presents methodological techniques like optimization approaches, qualitative approaches, and multi-criteria decision-making.
- Examines critical issues, including sustainability, ergonomics, quality assurance, and accuracy.
- Covers the Internet of Things–enabled smart systems and artificial intelligence in smart systems.
- Elaborates on management strategies and quality assurance frameworks for smart systems.

The text is primarily written for senior undergraduate/graduate students and academic researchers in the fields of mechanical engineering, manufacturing engineering, industrial engineering, production engineering, and aerospace engineering.

Dr. Faisal Talib is a Professor in the Department of Mechanical Engineering, Zakir Husain College of Engineering & Technology, Aligarh Muslim University (AMU), Aligarh, (U.P.), India. He holds a Ph.D. degree from IIT Roorkee and Masters in Industrial Engineering from AMU. He has over 26 years of teaching and research experience and has more than 115 publications to his credit in national/international journals and published 10 Book Chapters. He has successfully guided 04 PhDs. His work has been published and cited in various peer-reviewed refereed journals. Additionally, he is a member of various advisory boards/editorial boards of International Journals and serving as an expert reviewer in the review panel of several International Journals of repute. He is also a member of notable professional bodies some of which includes Indian Society for Technical Education (ISTE), The Institution of Engineers (IEI), International Association of Engineers (IAENG) and International Engineering and Technology Institute (IETI).

Dr. Muhammed Muaz is working as an assistant professor in the Department of Mechanical Engineering, Zakir Husain College of Engineering and Technology, Aligarh Muslim University (AMU), Aligarh, U.P., India. His research areas include metal cutting operations, friction stir welding, manufacturing automation, adaptive control of manufacturing processes, robotics, and automation, cutting fluids, minimum quantity lubrication, finite element method of machining, powder metallurgy, materials characterization, and optimization of manufacturing processes. He has published several research articles, including journal papers, conference papers, and book chapters. He has filled several patents on his own innovations. One of his patents is granted, and others are published. He is a reviewer of many refereed Q1 journals. He ranked first in B.Tech and second in M. Tech. He has completed his Ph.D. from IIT Kanpur.

Smart Systems
Methodological Approaches and Applications

Edited by

Faisal Talib and Muhammed Muaz

CRC Press is an imprint of the
Taylor & Francis Group, an **informa** business

Front cover image: greenbutterfly/Shutterstock

First edition published 2025
by CRC Press
2385 NW Executive Center Drive, Suite 320, Boca Raton FL 33431

and by CRC Press
4 Park Square, Milton Park, Abingdon, Oxon, OX14 4RN

CRC Press is an imprint of Taylor & Francis Group, LLC

ISBN: 978-1-032-46900-3 (hbk)
ISBN: 978-1-032-80077-6 (pbk)
ISBN: 978-1-003-49531-4 (ebk)

DOI: 10.1201/9781003495314

Typeset in Sabon
by Apex CoVantage, LLC

Contents

Preface

With the advent of interdisciplinary technological approaches and solutions, manufacturing industries as well as research institutes face challenges in integrating different technologies and diverse components into one system. This system approach calls for the design and manufacturing of smart systems with resilience, reconfigurability, and intelligence, resulting in the evolution of novel capabilities. These capabilities have significantly changed the manufacturing ecosystem by addressing environmental, societal, and economic challenges. The prime objective of this book is to provide a comprehensive and effective exchange of information on current developments and advancements in smart system applications. The book covers a broad range of topics about smart systems from a multidisciplinary point of view, starting with its critical review and continuing on smart tools, sustainable approaches, integration, Industry 4.0 aspects, and strategies in smart systems and their management. It also covers opportunities, challenges, and methodological approaches used by smart systems in the field of agriculture in order to enable growth toward smart agriculture context. It also focuses on knowledge transfer methodologies, smart design, and development in other innovative areas like biohumus, aroma culture, ergonomics, the pharmaceutical industry, and spiritual personality development. The book will be of interest to researchers, postgraduate students, academicians, practitioners, and engineers working in application areas such as agriculture, the automobile industry, healthcare, education, and others.

About the editors

Faisal Talib is a professor in the Department of Mechanical Engineering, Zakir Husain College of Engineering and Technology, Aligarh Muslim University (AMU), Aligarh, U.P., India. He holds a Ph.D. degree from IIT Roorkee and a masters in industrial engineering from AMU. He has over 27 years of teaching and research experience and more than 115 publications to his credit in national/international journals and conferences, including 45 papers in SCI/SCIE/Scopus journals and 8 book chapters. He serves as a research guide for Ph.D. scholars (four have successfully doctorate and three are in progress). His work has been published and cited in various peer-reviewed refereed journals, including the *Journal of Cleaner Production*, *Environmental Science and Pollution Research*, *Sustainability*, the *International Journal of Sustainable Engineering*, *Total Quality Management & Business Excellence*, the *International Journal of Health Planning and Management*, the *International Journal of Quality & Reliability Management*, the *International Journal of Sustainable Engineering*, the *International Journal of Service Business*, *Benchmarking: An International Journal*, the *International Journal of Productivity & Quality Management*, and the *International Journal of Mobile Communications*, to name a few. His special interests include Quality 4.0, total quality management (TQM), sustainable practices, operations management, multi-criteria decision-making (MCDM) techniques, and quantitative research. Additionally, he is a member of various advisory boards/editorial boards of international journals and serves as an expert reviewer in the review panel of several international journals of repute, including the *International Journal of E-Entrepreneurship and Innovation*, *International Journal of Productivity Management and Assessment Technologies*, *International Journal of Information Systems in the Service Sector*, *International Journal of Business Intelligence Research*, and *International Journal of Service Science, Management, Engineering, and Technology*. He is also a member of notable professional bodies, including the Indian Society for Technical Education (ISTE), The Institution of Engineers (IEI), the International Association of Engineers (IAENG), and the International

Engineering and Technology Institute (IETI). To conclude, he has more than 4400 citations (h-index = 31, i-10-index = 61) to his credit, indexed by Google Scholar.

Muhammed Muaz is working as an assistant professor in the Department of Mechanical Engineering, Zakir Husain College of Engineering and Technology, Aligarh Muslim University (AMU), Aligarh, U.P., India. His research areas include metal cutting operations, friction stir welding, manufacturing automation, adaptive control of manufacturing processes, robotics, and automation, cutting fluids, minimum quantity lubrication, finite element method of machining, powder metallurgy, materials characterization, and optimization of manufacturing processes. He has published several research articles, including journal papers, conference papers, and book chapters. He has filed two patents on his own innovations. One of his patents is granted, and one is published. He is a reviewer of many refereed Q1 journals. He is a guest editor of the special issue "Sustainable Developments & Innovations in Manufacturing" [Sustainability] (IF: 3.889, ISSN: 2071–1050). He teaches UG and PG courses. In addition to these courses, he is assigned the teaching load in various labs, including the manufacturing technology lab, hardware lab, computer programming lab, kinematics and stress analysis lab, engineering graphics lab. He ranked first in B.Tech and second in M. Tech. He has completed his Ph.D. from IIT Kanpur.

Contributors

A. Abdullah
Department of Mechanical Engineering
Zakir Husain College of Engineering and Technology
Aligarh Muslim University
Aligarh, Uttar Pradesh, India

Mohammad Ahmad
Department of Pharmacy
Integral University
Lucknow, India

Tufail Ahmad
Centre of Agricultural Education, Faculty of Agriculture
Aligarh Muslim University
Aligarh, Uttar Pradesh, India

Juber Akhtar
Department of Pharmacy
Integral University
Lucknow, India

Mohd. Aftab Alam
Department of Mechanical Engineering
Zakir Husain College of Engineering & Technology
Aligarh Muslim University
Aligarh, India

Mohammad Azad Alam
Interdisciplinary Research Centre for Sustainable Energy Systems
King Fahd University of Petroleum and Minerals
Dhahran, Saudi Arabia

Dr. Mohammed Ali
Department of Mechanical Engineering
Aligarh Muslim University
Aligarh, Uttar Pradesh, India

Mohammad Anas
Sharda School of Humanities and Social Sciences
Sharda University
Greater Noida, India

Kuzmin Anton
Department of Mechanization of Agricultural Products Processing
National Research Mordovia State University
Saransk, Russia

Imad Asif
Department of Mechanical Engineering, Zakir Husain

College of Engineering and Technology
Aligarh Muslim University
Aligarh, India

Mohammad Asjad
Department of Mechanical Engineering
Jamia Millia Islamia
New Delhi, India

Mohammad Azeem
Department of Mechanical Engineering, Faculty of Engineering
Universiti Technologi Petronas, Malaysia

Badruddeen
Department of Pharmacy
Integral University
Lucknow, India

Irfan Anjum Badruddin
Mechanical Engineering Department, College of Engineering
King Khalid University
Abha 61413, Saudi Arabia

Soumyabrata Banerjee
Amity Institute of Physiology and Allied Sciences
Amity University
NOIDA, U.P., India
Assistant Professor Department of Human Physiology
Vidyasagar University
Midnapore, West Bengal, India

Basab Chaudhuri
Heritage Institute of Technology
Kolkata, India

Azhar Equbal
Department of Mechanical Engineering
Jamia Millia Islamia
New Delhi, India

Sh.A. Eshkobilov
Department of Biotechnology
Tashkent Institute of Chemical Institute
Uzbekistan

Muhammad Farhan
Department of Mechanical Engineering
School of Science & Technology
Glocal University
Saharanpur, India

Zainab Fatima
Department of Computer Science and Engineering
Integral University
Lucknow, India
Space Application Centre
Indian Space Research Organization (ISRO)
Ahmadabad, India

Md. Mahedi Hasan
PhD Fellow, Department of Management Studies
University of Rajshahi
Bangladesh

Syed Mubashshir Hasan
Department of Civil Engineering
Zakir Husain College of Engineering and Technology
Aligarh Muslim University
Aligarh, India

Dr. Mohd. Shaaban Hussain
Department of Mechanical Engineering

Aligarh Muslim University
Aligarh, India

Sheeba Jilani
Department of Chemical Engineering
Aligarh Muslim University
Aligarh-202001, Uttar Pradesh, India

Md. Fazlul Karim
Department of Biochemistry
University of Calcutta
Kolkata, India

M. Masroor A. Khan
Department of Botany
Aligarh Muslim University
Aligarh, Uttar Pradesh, India

Mohammad Irfan Khan
Department of Pharmacy
Integral University
Lucknow, India

Mohd Usman Khan
Department of Computer Science and Engineering, Faculty of Engineering
Integral University
Lucknow, Uttar Pradesh, India

Sara Khan
Department of Pharmacy
Integral University
Lucknow, India

Zahid A. Khan
Department of Mechanical Engineering
Jamia Millia Islamia
New Delhi, India

Nortoji A. Khujamshukurov
Department of Biotechnology
Tashkent Institute of Chemical Institute
Research Institute of Plant Genetic Resources
Uzbekistan

Shara Khursheed
Department of Mechanical Engineering
Faculty of Engineering, Integral University
Lucknow, India

D.X. Kuchkarova
Tashkent University of Architecture and Civil Engineering
Uzbekistan

Mohan Kumar
Department of Chemistry
Shri Varshney College
Aligarh, India

Ghulam Mohiuddin
Department of Computer Application, Faculty of Engineering
Integral University
Lucknow, India

Muhammed Muaz
Department of Mechanical Engineering, Faculty of Engineering and Technology
Aligarh Muslim University
Uttar Pradesh, India

Mrinal Kanti Poddar
Department of Biochemistry
University of Calcutta
Kolkata—700 019, India

Emeritus Medical Scientist
Department of Pharmaceutical Technology
Jadavpur University
Kolkata, India

Md. Sohel Rana
College of Business Administration
IUBAT-International University of Business Agriculture and Technology
Dhaka, Bangladesh

Arsalan Sartaj
Department of Electrical Engineering
Zakir Husain College of Engineering and Technology
Aligarh Muslim University
Aligarh, India

Faisal Talib
Department of Mechanical Engineering
Faculty of Engineering and Technology
Aligarh Muslim University
Uttar Pradesh, India

Pankaj Kumar Tyagi
Leather and Footwear Technology
University Polytechnic, F/o Engineering and Technology
Aligarh Muslim University
Aligarh-202001, Uttar Pradesh India

Md. Shawan Uddin
Department of Management Studies
University of Rajshahi
Bangladesh

Muhammad Zaid
School of Industrial and Information Engineering
Politecnico Di Milano
Milan, Italy

Chapter 1

Management strategies in smart systems

Arsalan Sartaj, Syed Mubashshir Hasan, and Imad Asif

1.1 INTRODUCTION

With the increasing global competition, organizations need to scale up as we process in an economy with rapid technological advancements and globalization. To survive and thrive, organizations must prioritize strengthening the product innovation cycle as a key driver of their success. Innovation has become an important element to gain a competitive advantage and attain economic success. However, many organizations struggle to innovate due to a lack of practical management frameworks.

This chapter proposes steps to bridge the gap in current smart systems by introducing "management strategies." The focus of this chapter is to explore how innovation can be incorporated into organizations' routines by connecting the innovation ecosystem to their corporate strategy through smart systems. Smart systems require a combination of different management strategies to function optimally.

By utilizing smart systems, organizations can improvise, innovate their daily routines, and attain a competitive advantage in the market through extensive market penetration. This chapter will discuss how smart systems can be used to foster innovation, the different management strategies that can be used in smart systems, and how to connect the innovation ecosystem to firms' corporate strategy.

Managing smart systems requires a combination of different management strategies. No one strategy fits all industries, so it is important to make sure every industry is adapting to the framework. It can be done by controlling, monitoring, and optimizing these systems. This requires extensive knowledge of the whole system to make sure there are no errors in the optimization.

The goal of this chapter is to provide a guiding framework for organizations seeking to integrate changes into their daily routines by leveraging smart systems. The insights provided in this chapter will be useful to management executives to stay lean in a changing business environment.

DOI: 10.1201/9781003495314-1

1.2 NEED FOR MANAGEMENT STRATEGIES IN SMART SYSTEMS

The intricacy of managing smart systems might make it a difficult undertaking. To ensure their seamless, organized, and ideal operation, organizations must create efficient management techniques. So, there is a need to use a variety of management techniques, including but not limited to data, network, security, performance, resource management, and strategic planning.

Since it is crucial to successfully gather, store, analyze, and visualize data, it makes data management a key component of smart systems. A lot of data is produced by smart systems. These activities are crucial for firms' ability to manage this data effectively and derive valuable insights. The bulk of datasets when processed and analyzed uses different approaches, including data mining, machine learning (ML), and artificial intelligence.

Data is typically collected from different sources, including sensors, cameras, and other devices, in smart systems. The information gathered must be precise, timely, and pertinent to the system in question. The accuracy of the data is important for the adoption of smart systems. Incorrect judgments can be made as a result of low-accuracy data, which can compromise the stability and functionality of the system.

Companies must ensure they have the infrastructure and resources required to process the massive amounts of data produced by intelligent systems. This covers the amount of storage required to store the data, various tools for visualizing it, and methods for processing the data collected using data cleaning and data quality management.

Organizations must also make sure that the data they acquire is securely preserved and free of cyber-threat vulnerabilities. Prevention of data exploitation by compromised actors is done by adding security mechanisms, including firewalls, intrusion detection systems, and encryption, which must be put in place. Cyberattacks have impacted intellectual property in a number of situations; therefore, it is critical to ensure that this doesn't happen going forward, as we rely on intelligent systems.

In order for enterprises to gain insights from the data produced by intelligent technologies, effective data management is crucial. Organizations can perform trend analysis on the data, spot patterns and abnormalities, and use this information to guide their judgments about how well the system is working. By detecting problem areas and taking appropriate action, companies can use effective data management to enhance system performance.

Network management is an important aspect of managing smart systems, as it involves managing the infrastructure that connects the smart systems. The network infrastructure consists of hardware and software peripherals that allow communication and data exchange between devices within the smart system. It encompasses several activities, including monitoring the network and ensuring smooth operation, identifying and resolving any issues, and optimizing the network for performance. Optimized networks have lowered costs and take less bandwidth of resources.

To effectively manage a network, organizations need to ensure that their network infrastructure is scalable, reliable, robust, and secure. These are key metrics for meeting projections and the adoption of the smart system. It ensures that the network infrastructure is robust at high traffic volumes. At the same time, reliability ensures that the network is available and accessible to users at all times. The primary metric for adoption is the ability of the network to be reliable and safe. Safe networks build trust in users.

Organizations need to ensure network security and implement security protocols, such as firewalls, intrusion detection systems, and encryption, to avoid data being compromised.

To understand effective management, we need to understand different parameters like network topology, traffic patterns, and potential security threats. Organizations can use network management tools and software to evaluate performance benchmarks. Implementation of these strategies will assure efficiency, reliability, and security, leading to a better user interface, streamlined operations, and seamless flow.

Security management is an essential aspect of managing smart systems. The breach of data is a common issue that occurs in smart systems. In order to mitigate these threats, organizations must implement robust security procedures to ensure the confidentiality of data.

One of the major security procedures that allow us to do the same is firewalls. Firewalls prevent unauthorized access to the network by blocking incoming and outgoing traffic based on predefined rules that are determined by users. The users or admin gets to block certain traffic based on malicious behavior. One must ensure that their smart system has an active firewall system in place.

Another security protocol commonly used in managing smart systems is intrusion detection systems (IDSs). IDS helps detect and respond to potential security threats by monitoring network traffic and identifying suspicious activity. It can be either host-based or network-based, and organizations must choose the appropriate type based on their smart system's needs.

Encryption is another critical security protocol used in managing smart systems. It helps protect data by converting it into an unreadable format by cipher or hash, which can only be deciphered using a decryption key. Encryption is commonly used to protect data in transit or data at rest, and organizations must ensure that their smart system has adequate encryption protocols in place.

Organizations must also ensure that their systems' security is regularly updated and maintained to mitigate evolving threats. This involves regularly updating software and firmware, patching vulnerabilities, and ensuring that security protocols are up-to-date.

Security management is a crucial aspect of managing smart systems. Organizations must implement robust security protocols such as firewalls, intrusion detection systems, and encryption to ensure the confidentiality, integrity, and availability of data. Additionally, regular updates and maintenance are necessary to mitigate evolving cyber threats.

Effective performance management is another important factor. It involves monitoring the system's performance, identifying potential issues or bottlenecks, followed by root cause analysis to ensure identification of the problem, and taking several actions to optimize system performance. Organizations need to ensure that their smart systems are performing at an optimal level to achieve the desired outcomes.

To effectively manage the performance of a smart system, organizations need to establish key performance indicators (KPIs) that align with their system's goals and objectives. These KPIs can be used to measure the system's performance and identify areas where improvements are needed. KPIs allow managers to show potential projections and track the growth of the strategy, which in turn assures that no time is wasted on any ineffective strategies.

It is also important to continuously monitor and analyze the system's performance over time relative to benchmarks, so further optimization can be achieved. This can involve using predictive analytics and ML algorithms to identify performance issues, allowing for mitigation and optimization of the system.

Resource management involves identifying the resources required to operate the smart system effectively, such as hardware, software, and personnel, and managing them to ensure their availability when needed. Organizations need to ensure that they have the necessary resources to operate and establish the smooth functioning of the system.

Strategic planning is another critical component of managing smart systems. It involves the identification of objectives, goals, and projected outcomes. This is followed by an action plan roadmap with different action items that can be followed. Organizations must ensure that their strategic plan is in sync with their overall corporate strategy to maximize benefits.

Organizations need to identify the most effective plan with the help of KPIs and other suggested methodologies for their specific smart system by assessing industry needs, goals, and deliverables of the system. Effective management can provide several benefits, including improved efficiency, reduced costs, and enhanced quality of products and services.

1.2.1 Different components of management strategy

A management strategy typically consists of various key components that allow executives to reach the desired outcomes. Some of the essential components are discussed here.

1.2.1.1 Goal setting

Clear and precise goal setting and objectives are fundamental to guiding an effective management strategy. They set the direction for the organization and provide a clear direction for the growth of the system. Goals should be tangible so that it is possible to build them with KPIs in mind and achievable targets.

To be effective, goals and objectives should be SMART: Specific, Measurable, Achievable, Relevant, and Time-bound. Specific goals define action items to be achieved, while measurable goals provide a quantifiable metric for evaluating progress toward achieving them. Achievable goals ensure that they can be accomplished with the given resources at hand, while relevant goals are aligned with the organization's overall mission and vision. Time-bound goals have a defined timeline for completion, ensuring that they are completed within a set timeframe.

Having SMART goals and objectives in place ensures that the management strategy is focused, cohesive, and impactful. A clear set of goals and objectives provides a roadmap for the management strategy, helping organizations to allocate resources effectively and prioritize actions that are aligned with achieving desired outcomes. It also enables stakeholders to monitor progress toward achieving these goals and adjust the management strategy accordingly.

1.2.1.2 SWOT analysis

The SWOT (Strengths, Weaknesses, Opportunities, and Threats) analysis is a valuable tool for strategy development. It facilitates the identification of internal and external factors that govern the success of the strategy, such as emerging technologies, market trends, and changes in regulatory frameworks. The SWOT analysis assists in generating a comprehensive understanding of the smart system's current state and its potential opportunities for growth and improvement.

By conducting a SWOT analysis, organizations can identify their strengths and weaknesses, such as a highly skilled workforce, intellectual property, or inadequate funding. They can also assess opportunities that the smart system can capitalize on, such as market trends or emerging technologies. Moreover, a SWOT analysis helps to give a bird's-eye view of the system and help prioritize the different tasks, as stated in Figure 1.1.

This analysis provides a structured framework for decision-making, resource allocation, and risk management. It helps organizations allocate resources effectively by focusing on the areas that offer the most significant opportunities for growth while minimizing risks. Additionally, it aids in the identification of potential risks and the development of mitigation plans to minimize their impact.

	Strengths (S)	*Weaknesses (W)*
Opportunities (O)	SO strategies Leverage strengths to maximize opportunities Attacking strategy	WO strategies Counter weaknesses through
Threats (T)	ST strategies Leverage strengths to minimize threats Defensive strategy	WT strategies Counter weaknesses and threats Build strengths for defensive strategy

Figure 1.1 SWOT Analysis Framework.

1.2.1.3 Action plan

To effectively implement a management strategy for smart systems, an action plan is necessary. It serves as a detailed roadmap that outlines the steps necessary to achieve the strategy's goals and defines the clear priority tasks in chronology. It includes specific tasks that need to be done, timelines for each action, responsibilities assigned to team members, and milestones for tracking progress.

The action plan serves as a blueprint for execution and helps to ensure that the alignment of efforts is there. It prevents missing out on the macro impacts on the whole value chain. It helps make the information consumable and easy to understand for smart system users. It enables decisions backed by KPIs and objective and Key results (OKRs) with active tracking of these metrics.

Strategy deployment requires careful consideration of the objectives and the resources required to achieve any curated OKRs. It is important to ensure that the plan is attainable, alongside managing deadlines and other metrics. Regular monitoring and updating of the action plan are also critical to ensuring that it remains relevant and effective.

Smart systems require ongoing monitoring and evaluation to determine whether management strategies are effective in achieving the desired outcomes. Daily analytical evaluation of the system's performance can help spot potential issues and areas for improvement. This process can include the evaluation of KPIs to measure progress toward achieving specific goals and objectives. Furthermore, if any changes to KPIs are to be factored in, regular assessment aids in that as well.

Moreover, periodic evaluation can provide organizations with valuable insights into the system's behavior, allowing for the early detection of potential problems. Iterations done prevent any failures in smart systems so that downtime and other disruptions that may negatively impact organizations' operations are avoided. Continuous evaluations allow for adjustments to be made to the management strategy wherever required and to battle test the strategies and actively iterate.

1.2.1.4 Resource allocation

To effectively manage smart systems, it is crucial to allocate resources appropriately to ensure the success of the management strategy. It encompasses a wide range of resources, including capital, human resources, technology, and time. This allows proper utilization of the resources with no capital inefficiencies.

Capital here refers to financial resources that are necessary to invest. The investment allows units to scale, and that leads to the adoption of systems. Human resources serve an important role in the ecosystem, as skilled professionals are needed to design, build, and operate the smart system. A lot of resources go into the development of smart systems; hence, the need for proper resource allocation is important.

Technologies such as hardware and software play a monumental role as well. Proper allocation of these resources is necessary to ensure optimal system performance. Time is another critical resource that needs to be managed effectively to ensure that the project timelines and deadlines are met. This can be ensured by using several project management frameworks.

It is important to note that resource allocation must be aligned and coordinated. The resources must be allocated in a way that maximizes the utility and reduces any extra operational expenditure. This requires careful planning, analysis, and decision-making to ensure that the resources are used effectively and efficiently.

1.2.1.5 Communication

Effective communication is critical to ensure that everyone involved in the strategy understands its goals, objectives, and action plan. Communication is an iterative process that is supposedly a two-way process. Active feedback and input from all stakeholders are always encouraged and allow clearer idea propagation.

Effective communication can also help promote ideas, sort relevancy, and clearer propagation. Customers who are kept informed of changes or new features in the smart system are more likely to have positive experiences, leading to increased satisfaction and loyalty.

In addition, communication should be tailored to the specific needs of each stakeholder. For instance, the communication approach of customers may differ from that of employees or suppliers. The method of communication, such as email, meetings, or reports, should also be chosen based on its suitability for the audience and the type of information being conveyed.

Finally, regular communication is essential to ensure that stakeholders are kept informed of progress toward achieving the strategy's goals and objectives. This allows for adjustments to be made if necessary, ensuring that the smart system is on track to deliver the desired outcomes.

1.2.1.6 Continuous improvement

It is crucial to review and refine the strategy regularly. Continuous iterations involve a structured process of evaluating the strategy's effectiveness, identifying areas for improvement, and implementing changes based on feedback from stakeholders. The lean methodology can be referred to as where active iterations are done by factoring in consumer feedback. This process requires a culture of continuous learning, a willingness to adapt, and a commitment to staying up-to-date with the latest technologies and best practices.

To support continuous improvement, organizations must establish a feedback loop to collect and analyze data on the strategy's performance. This data can be used to identify potential areas for improvement and inform decision-making [1]. Additionally, organizations must mobilize stakeholders to actively make suggestions for improvement. This feedback can be collected through several feedback channels like forms, sessions, and A/B testing.

Continuous improvement is an ongoing process that requires consistency and active effort. Passively managing the strategies would just lead to a poor product stack. It involves regular reviews of the strategy, ongoing learning and development, and a willingness to embrace change. By prioritizing continuous improvement, organizations can ensure that their management strategy remains effective and relevant in an ever-changing landscape of smart systems [2].

1.2.1.7 Innovation

Innovation is a crucial component of any successful management strategy for smart systems. The primary goal of innovation is to create new and better products, processes, or services that meet the changing needs of customers and improve the overall performance of the organization. As aforementioned, the product-innovation cycle ensures the longevity of a product's competitive advantage; if the system keeps iterating, it emerges as a leader in the markets.

The objectives of an innovation strategy are to increase revenue, enhance customer satisfaction, and improve market share. Innovating actively leads to a longer product innovation cycle that helps the product gain a first-mover advantage in the market; this strategy in itself leads to more market penetration and adoption.

As aforementioned, an action plan is necessary that outlines the specific steps needed to achieve the desired outcomes of the innovation strategy. This includes identifying potential new prospects, extensive market research, low-fidelity or high-fidelity prototypes, testing, and iterations of the final product.

Monitoring and evaluation are critical components of any successful innovation strategy. This involves tracking KPIs and analyzing feedback from customers and stakeholders to ensure that the new product or service is meeting its objectives.

Continuous improvement is essential to maintain the competitive advantage gained through innovation. The organization should consistently review its processes and outcomes and seek ways to improve and optimize the innovation process [3, 4].

Management strategies are applied to achieve the desired outcome by using the aforementioned steps to implement an innovation strategy. By using data-driven decision-making and leveraging advanced technologies such as artificial intelligence and ML, organizations can identify new opportunities for innovation, actively innovate, and gain a first-move advantage in the market.

1.2.1.8 Financial performance

Financial performance is a critical component of any organization's success, and smart systems can play a vital role in achieving financial goals. The goal of financial performance is to maximize monetary returns while reducing operational costs, which leads to higher efficiency and resource utility. To achieve this goal, organizations must set clear objectives, allocate appropriate resources, develop action plans, monitor and evaluate progress, and continuously improve their strategies.

Smart systems can help organizations achieve these objectives by training the models using modern techniques, making predictions for future use cases, and actively managing the situations. For example, a retail organization can use smart systems to analyze sales data, identify trends, and adjust pricing strategies accordingly. Another great example is modern devices like smart grids used in power systems that allow active load management during peak loads based on data analysis [5].

To apply management strategies for smart systems to financial performance, organizations must first define their financial goals and objectives. They should then allocate appropriate resources to develop and implement a smart system that can support these goals.

Next, organizations should develop an action plan that outlines specific tasks, timelines, and responsibilities. They should also establish a monitoring and evaluation process to track progress and identify areas for improvement.

Continuous improvement is key to achieving financial performance goals. Organizations should regularly review and refine their strategies, leveraging smart systems to identify areas for optimization.

In summary, applying management strategies for smart systems in financial performance can help organizations maximize revenue, minimize costs, and achieve a healthy bottom line. By setting clear goals and objectives, allocating appropriate resources, developing action plans, monitoring and evaluating progress, and continuously improving strategies, organizations can leverage the power of smart systems to achieve their financial goals [6].

1.2.1.9 Risk management

Risk management is an essential component of any organization, and it involves the identification, assessment, prioritization, and mitigation of risks that could negatively impact organizations' objectives. An important part of risk management is to come up with mitigation strategies. In the context of smart systems, it becomes even more critical as these systems are complex, interconnected, and often rely on data-driven decision-making.

To effectively manage risks in smart systems, it is important to understand the value chain of the system. This involves identifying the potential risks in the chain and their impact on organizations' goals, as well as the resources required to mitigate those risks.

A strategy should be developed that outlines the steps required to manage the identified risks. This plan should be regularly monitored and evaluated to ensure that it is effective and that any iterations are to be made.

Continuous improvement is essential in managing risks associated with smart systems. This involves regularly reviewing and updating the risk management strategy to ensure that it remains relevant and effective. Threats opportunities weaknesses and strengths analysis is an extension of SWOT analysis that helps us identify and mitigate risks.

Overall, a management strategy that includes effective risk management can help organizations achieve their objectives while minimizing potential risks. By identifying, assessing, and mitigating risks, organizations can ensure that their smart systems operate efficiently and effectively, leading to improved results and smoother operations.

1.2.1.10 Sustainability

Sustainability is the process of carrying out tasks without hindering the future scope of work. It is a critical component of any business strategy today. In order to achieve sustainability goals, a comprehensive management strategy is needed. The goals of a sustainability strategy may include reducing carbon footprint, conserving resources, reducing waste, promoting environmental responsibility, and creating a more equitable usage of available resources. At present, we see many businesses adopting sustainable practices and strategies. The objectives of a sustainability strategy should be SMART. Resource allocation again plays a vital role, as it involves identifying the necessary resources and allocating them effectively for a larger duration without hindering short-term operations.

A sustainability projection plan that outlines the specific steps that need to be taken to achieve sustainability goals and objectives must be developed. Monitoring and evaluation are essential to track progress, identify areas for improvement, and ensure that the desired outcomes are being achieved. Impact analysis frameworks allow for measuring sustainability goals more carefully and are promoted as they are beneficial to the environment as well.

Management strategies for smart systems can be applied to sustainability to achieve the desired outcome. Smart systems can be used to monitor energy usage, track resource consumption, and optimize operational processes to reduce waste and promote sustainability. By leveraging data analytics and ML algorithms, smart systems can help organizations identify areas for improvement and optimize their resource utilization.

In summary, a comprehensive management strategy for sustainability that leverages smart systems can help organizations achieve their sustainability goals and objectives while optimizing resource utilization and promoting continuous improvement [7].

These are some of the components from which use cases can be explored for management strategies. It is important to conduct a thorough analysis of organizations' goals, challenges, and opportunities to identify the most relevant use cases for the management strategy.

1.2.2 Use cases

Use cases are specific scenarios or situations where the management strategy can be applied to achieve a desired outcome. The components of use cases might vary based on industries and demands. Use cases are tailored to specific demands, requiring an assessment of the industry. The following are some examples of components from which use cases can be explored.

1.2.2.1 Operational efficiency

Operational efficiency is a critical component of any management strategy for smart systems. It involves optimizing processes and workflows to maximize output while minimizing resource consumption. The goal of operational efficiency is to reduce waste, increase productivity, and ultimately improve the bottom line.

To achieve operational efficiency, managers need to define, assess, and develop a detailed strategy and monitor and evaluate performance on an ongoing basis. Continuous improvement is also essential, as the eventual goal is to continuously refine and optimize operations over time.

Consider a manufacturing plant that produces consumer goods. The plant management team wants to improve operational efficiency by reducing miscellaneous losses, increasing productivity, and decreasing the resource waste input-to-output ratio. The team identifies a number of potential strategies, including implementing lean manufacturing principles, optimizing production schedules, and investing in new technology.

After evaluating each strategy, the team decides to implement lean manufacturing principles. They define specific goals and objectives, including reducing waste by 20%, increasing production output by 15%, and improving overall quality by 10%. They allocate resources to hire a lean manufacturing consultant and train staff on the new processes. A good manager focuses on all the must-haves and deprioritizes any extra elements that can be let go of to make this process leaner.

As a result of these efforts, the plant is able to achieve its goals and objectives, and the management team is able to demonstrate significant improvements in operational efficiency. By applying a management strategy focused on operational efficiency, optimized performance of the plant is achieved by reducing waste, increasing productivity, and improving overall profitability.

1.2.2.2 Customer experience

Smart systems allow lead management and also improve overall customer support experience. This has been achieved due to the presence of advanced systems that can handle high-volume data. This has become a crucial aspect of any organization's success. To achieve the desired outcome of providing an exceptional customer experience, a management strategy needs to be put in place.

The goal of the customer experience management strategy is to create positive interactions with customers throughout their journey with the organization. The objective is to increase customer satisfaction, loyalty, and retention, which will ultimately result in increased revenue and profitability.

Resource allocation should be focused on training employees, implementing technology solutions, and creating a feedback loop for customers. Action plans should include mapping the customer journey, identifying pain points, and addressing them with the appropriate resources in the various phases, as given in Figure 1.2.

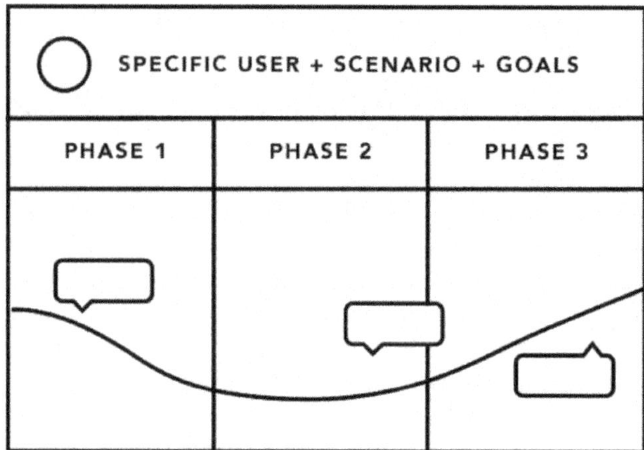

Figure 1.2 Customer Journey Map.

Monitoring and evaluation of the customer experience is an active process that requires a consistent system bandwidth, thus making automation crucial. Data on customer experience will help identify areas of improvement and determine the effectiveness of the action plan.

It is critical to the success of the customer experience management strategy. Organizations should regularly review and refine their approach based on feedback from customers and employees to ensure that smart systems are adaptive and iterative to stay relevant in the market.

The management strategy in the case of smart systems can be applied to achieve the desired outcomes such as leveraging technology solutions—for example, chatbots, AI-powered customer service, and personalized recommendations. These tools can decrease the time load on company resources, remove any substantial biases in the system, log data more accurately, and allow the creation of a standard operating procedure (SOP).

1.2.2.3 Talent management

Talent management is a critical component of any organization, especially those utilizing smart systems. As smart systems lie on the emerging technology spectrum, it is important for organizations to involve the best human resources available in the market as it is at the forefront of innovation and new developments. In this use case, we will consider a hypothetical software development company that specializes in creating smart systems for various industries.

The goal of this organization is to develop innovative smart systems that meet the needs of its clients and stay ahead of its competitors. The objective is to attract and retain the best talent in the industry. Organizations have to make sure that they are up-to-date with the best practices in the industry so as to retain the best talent. It is important to understand that for larger organizations it is expensive to do multiple hiring rounds. It is also important that organizations provide competitive compensation and benefit packages to employees.

A plan could include implementing a robust recruitment process that attracts top talent from universities, industry events, and other sources. Additionally, creating a strong employer brand and employee value proposition can help organizations stand out and attract or retain the best candidates.

To monitor and evaluate the talent management strategy, organizations can conduct regular performance reviews, track employee turnover rates, and gather feedback from employees through surveys and other communication channels. Talent management may include high-resource strategies like ERPs and HRM software that may involve, but are not limited to, payroll management, international remittance, and many more.

Overall, the smart systems management strategy can be applied to talent management to achieve the desired outcome of attracting and retaining the best talent in the industry. By aligning organizational goals and objectives with the skills and abilities of its workforce, investing in the right resources, and continuously improving the talent management strategy, the organization can maintain a competitive edge in the smart systems market [8].

1.3 CONCLUSION

In the era of Industry 5.0 and the Internet of Things, smart systems play a crucial role in organizations. In order to ensure that these smart systems are implemented seamlessly, effective management strategies need to be deployed. It is becoming increasingly important for businesses to innovate their workflows to remain relevant. In this economic landscape, practical management frameworks gain primacy in aiding organizations.

Effective management strategies are essential for the successful implementation of smart systems in organizations. Businesses need to prioritize innovation to remain competitive. They have to implement innovative approaches to their workflow to achieve economic success. However, the lack of practical management frameworks has made it difficult for many organizations to effectively incorporate innovation into their daily routine.

This chapter proposed "Management Strategies in Smart Systems" as a solution to bridge this gap. By connecting the innovation ecosystem to the firm's corporate strategy through the use of smart systems, organizations can effectively control, monitor, and optimize these complex systems.

The different management strategies, including data, network, security, performance, and resource management, were explored in detail, highlighting their importance in the effective functioning of smart systems. The most effective strategy in different use cases was also discussed, emphasizing the need for a tailored approach to management.

Moreover, the integration of technical expertise, data analytics, and strategic planning was identified as a critical component for further maximizing the performance of smart systems. Organizations must also consider the ethical and social implications of implementing smart systems, such as the impact on employment and privacy.

Overall, the adoption of management strategies in smart systems can provide organizations with a competitive edge in today's rapidly changing business landscape. Businesses must embrace innovation and leverage technology to improve their operations and ultimately achieve success in the long term.

REFERENCES

[1] Barton, R., & Thomas, A. (2009). Implementation of intelligent systems, enabling integration of SMEs to high-value supply chain networks. *Engineering Applications of Artificial Intelligence*, 22(6), 929–938. https://doi.org/10.1016/j.engappai.2008.10.016

[2] Blanco, I., Cenni, F., Carminati, R., Ciccazzo, A., Dalle Feste, S., Fummi, F., . . . Roselli, G. (2016). Smart system case studies. *Smart Systems Integration and Simulation*, 195–227. https://doi.org/10.1007/978-3-319-27392-1_8

[3] Schoitsch, E. (2019). Beyond smart systems—creating a society of the future (5.0) resolving disruptive changes and social challenges. *Proceedings of IDIMT 2019, Innovation and Transformation in a Digital World*, 450. https://doi.org/10.5281/zenodo.3605686

[4] Ho, J.-Y., & O'Sullivan, E. (2017). Strategic standardisation of smart systems: A roadmapping process in support of innovation. *Technological Forecasting and Social Change*, 115, 301–312. https://doi.org/10.1016/j.techfore.2016.04.014

[5] Shaban, A., Farhan, M., & Ahmed, S. (2022). Building a smart system for preservation of government records in digital form. In *2022 international congress on human-computer interaction, optimization and robotic applications (HORA)* (pp. 1–6). IEEE. https://doi.org/10.1109/HORA55278.2022.9800034

[6] Jones, A., Subrahmanian, E., Hamins, A., & Grant, C. (2015). Humans' critical role in smart systems: A smart firefighting example. *IEEE Internet Computing*, 19(3), 28–31. https://doi.org/10.1109/mic.2015.54

[7] Romero, M., Guédria, W., Panetto, H., & Barafort, B. (2020). Towards a characterisation of smart systems: A systematic literature review. *Computers in Industry*, 120, 103224. https://doi.org/10.1016/j.compind.2020.103224

[8] Yamazaki, Y., & Maeda, J. (1998). The SMART system: An integrated application of automation and information technology in production process. *Computers in Industry*, 35(1), 87–99. https://doi.org/10.1016/s0166-3615(97)00086-9

Chapter 2

A critical review on smart manufacturing

Zahid A. Khan and Irfan Anjum Badruddin

2.1 INTRODUCTION: BACKGROUND AND DRIVING FORCES

Manufacturing is the process of converting the raw material into the finished product with the help of a man, a machine, and tools. The substance engaged in this process is also transformed chemically and biologically (Moldavska & Torgeir 2017). The main goal of the production system is to use technical expertise in the design and optimization of the process. It is one of the key sectors that contributes significantly to the national economy (Kennedy 2010). The generated final goods are sent to suppliers or distributors, where they are purchased by customers and sold. Traditional manufacturing techniques have a much longer lead time and a longer production cycle (Equbal et al. 2015). In addition to having little to no customization options, little to no design flexibility, and little to no automation, they also get the most out of the physical work needed. For instance, issues with undercuts, draft angles, and thickness homogeneity may arise if a mold is made. Moreover, manufacturing costs are often considerable; thus, the aim may be met by increasing production volume, which will lead to a very cheap unit cost. Also, the gap between the creation of goods and their availability on the market is too wide, which hurts both the seller and the buyer. Population growth and globalization have increased the pace of production consumption, which has increased provider competitiveness for quick manufacturing and a quick supply chain system. In order to cut production costs, increase early shipping of production units, and shorten production time, companies are using unique processes and new technologies.

One of them is smart manufacturing (SM), which offers a perfect way to get beyond these obstacles. Computers are incorporated into the manufacturing facility in this instance, and the production system is geared toward more automation and less human involvement (Cioffi et al. 2020). Artificial intelligence (AI), cutting-edge data analytics, always-on connection, input sensors, and computing capacity are all integrated into the conventional manufacturing process in an SM system (Lu et al. 2016). With greater degrees of design flexibility, quick design modifications, adaptability, and

 DOI: 10.1201/9781003495314-2

digital edge technology, SM technology helps shorten manufacturing times. With the help of this technology, it may be possible to achieve reliable and recyclable products with high production rates and an optimized supply chain system. Smart technologies enable modeling and simulation to be done before real production, reducing or eliminating the possibility of mistakes (Ang et al. 2017). The production system may be mechanized, and effective cyber security is also feasible. The implementation of a smart manufacturing system in a manufacturing facility is shown in Figure 2.1, where numerous smart tools are combined with the current production process to make the facility smart. The conventional factory is being equipped with tools like an AI-based computer vision system, robotic aided activities like robotic welding, robot assembly of car components, etc., additive manufacturing (AM) systems, modeling and simulations, and bid data technologies. Computers are taught to capture and propose the data from photos and videos in an AI-based computer vision system. AI models were applied to photographs that were recorded, which helped the computers categorize the items and react, such as unlocking iPhones using a facial recognition system (Lee et al. 2018). Robotics allows for more flexibility and exact outcomes in industrial processes like welding and assembling automotive components (ElMaraghy 2005). Every component may be built via AM, sometimes referred to as digital manufacturing (Equbal et al. 2022). Even a skyscraper can now be built utilizing AM technology in a matter of days. The use of modeling and simulation technologies in the industrial sector reduces production obstacles and results in more precise and adaptable products. A virtual model of a real-world item or process may be created using big data, making it simple to examine and make adjustments. The use of all these cutting-edge

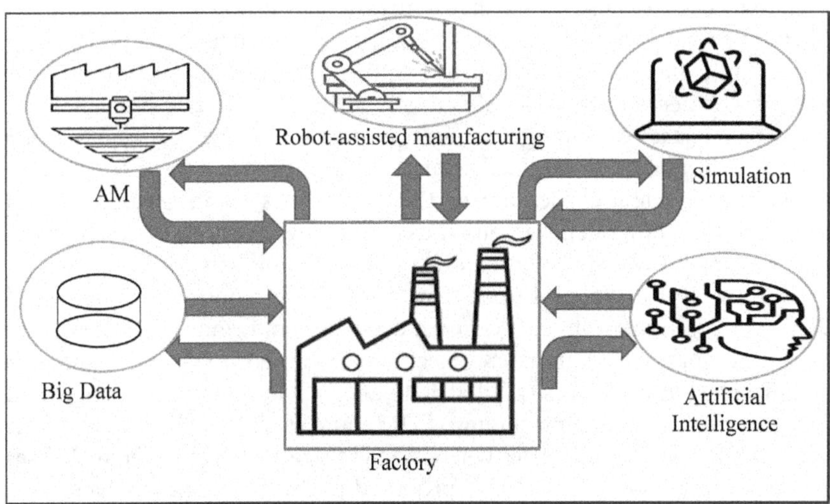

Figure 2.1 Smart Factory.

technology makes it simple to convert manual processes into intelligent, automated ones.

The current state of the industry shows that smart manufacturing technologies are highly sought after because of the many advantages and automation they provide. So, a review study that motivates researchers and practitioners to use it is necessary. So, the authors are encouraged to give a critical analysis of smart manufacturing and its use in the industrial setting. Hence, they present key smart manufacturing technologies and associated research, compiled from Web of Science, Scopus, and Google Scholar. The chapter is structured as follows: introduction, state of the art, discussion, conclusion, and limitations.

2.2 STATE OF THE ART

Smart manufacturing systems (SMS) place an emphasis on modern production processes and equipment that are enhanced by information technology integration. With real-time monitoring, flexible production, quick reaction to market changes and variations, cutting-edge sensors, big data analytics, and improved efficiency, every component of the manufacturing systems is digitalized in SMS (Lu et al. 2016). Important smart tools are addressed in this section, along with relevant research.

2.2.1 Additive manufacturing

The process of AM, sometimes referred to as 3D printing, involves creating a computer-aided design (CAD) design of the component, which is then loaded into 3D software that is already pre-installed on the computer. Slicing or tessellation of the part is carried out depending on the height of the component to be manufactured (Equbal et al. 2013). According to the AM process, either the material is extruded from the nozzle onto the build platform or laser technology is employed to fuse powder deposited on the platform (Equbal et al. 2022). The material deposited in a layer corresponds to the slicing height chosen during the process' design phase (Equbal et al. 2013) as the component is constructed layer by layer. The AM approach was automated with little human involvement thanks to computer integration, allowing for the consistent fabrication of exact models (Equbal et al. 2013). Stereolithography (SLA), fused deposition modeling (FDM), selective laser sintering (SLS), laminated object manufacture (LOM), and others are significant AM techniques (Equbal et al. 2022). Automobiles, casting, welding, electrical discharge machining (EDM), and several medical applications are just a few of the industries where AM is employed (Equbal et al. 2021a). In their study, Gao et al. (2022) proposed a novel strategy for facilitating the production of complex-shaped metal parts: incorporating AM into the casting procedure. They suggested segmenting intricate casting frameworks into several aspects. First, the casting substrate was created, then using AM, complicated features were created on the created casting substrate. This research sheds light on the integration of AM with casting and other techniques and offers a fresh approach. Wax patterns were swapped out for plastic components made using AM's fused deposition technique by Singh et al. (Singh et al. 2021). They concluded that shell cracking was an issue with plastic patterns that occurred during the burnout phase of the pattern. They decided that a wax sprue with a plastic design was the best way to prevent the issue and came to the conclusion that shell cracking was no longer a concern. Metal components are also produced via AM; however, concerns like flaws and poor quality are often encountered. To solve the problem, Cannizzaro et al. (2021) used a computer-based vision system and machine learning techniques. They suggested layer-by-layer metal component production with in-situ camera-based monitoring and flaw identification. Images acquired during the manufacturing process are gathered using a camera that is positioned on top of the machine. A generative adversarial network (GAN) is used to assess these photos in real time for defects, and the process is automated as a result. In order to create dissimilar welds on aluminum and commercially pure titanium, Sridharan et al. (2016) employed an ultrasonic-assisted AM method. They then examined how the microstructure and texture changed during the welding process. Their findings demonstrated that the ultrasonic AM-achieved aluminum texture at the interface is comparable to that produced by the accumulative roll bonding of aluminum alloys. They suggested that significant shear deformation at the interface serves as the primary basis for the bond formation process in ultrasonic AM.

A research was done by Equbal et al. (2019) to see whether the FDM method might be used to make electrodes for EDM. They used FDM to create plastic pieces for their investigation, which were then metalized to turn them into electrodes. The produced FDM electrode was used for EDM, and its performance in terms of material removal rate (MRR), tool wear rate (TWR), surface roughness (SR), and dimensional accuracy was compared with that of a traditional copper electrode (DA). The results demonstrated that, in terms of MRR, TWR, and SR, the machining performance of the FDM electrode was equal to that of the copper electrode. Nevertheless, owing to restrictions encountered during the metallization stage of electrode production, the DA generated by an FDM electrode is less effective. Another AM tool is fast tooling, which combines traditional tooling methods with AM techniques to create quick molds in production. Azhar et al.'s (Equbal et al. 2015) proposal to employ rapid tooling (RT) to effect a significant change in industry practice was included in this approach. Quick tooling may shorten the lead time for producing better-quality tools and ease the industrial revolution's bottleneck. They came to the conclusion that the rate of tool development across sectors was dramatically accelerating thanks to RT technology. The use of a tool electrode produced by selective laser melting (SLM) was recommended by Ablyaz et al. (2022) to produce a textured surface on the high-alloy steel 15Cr12H2MoWVNNB. In addition to modeling electrodes, the study of internal stresses produced during the SLM and electrode manufacturing processes is also taken into account. By simulating the electrode, they discovered that the corners of the tool electrode experience the highest internal stresses. This information allowed them to alter the support to further bind the portion to the machine substrate. They came to the conclusion that SLM can produce an electrode with a complicated profile and the necessary end surface arrangement. AM approaches for complicated medical treatments have shown to be creative and user-friendly. Equbal et al. (2020) provided insights on the application of AM in nephrology for teaching and surgical planning in this area. They showed how AM technology may be utilized to design and create urological devices, model urological procedures, schedule surgeries, and instruct patients and apprentices. They also said that the customization of medical-related items has improved steadily as a result of the advancement of AM technology. No one is unfamiliar with the terror of COVID-19. Each country's economy has been severely impacted by this virus, which has driven all of the countries to its knees. AM technology has proven to be a scientific boon in the fight against this epidemic. During the pandemic, when personal protective equipment was in limited supply, AM aided in the quick production and distribution of 3D face masks, face shields, hand sanitizer containers, hand gloves, AM-printed gowns, etc. (Equbal et al. 2021a). A reflective research on the utilization of AM techniques in medical applications was offered by Salmi (Salmi 2021). They argue that AM techniques are frequently used in the medical industry. They showed that sheet lamination is employed to create medical models or phantoms, whereas direct energy deposition is used to create bodily implants. All categories make use of powder bed fusion, material extrusion, and VAT photo polymerization. Thermoplastics, photopolymers, and metals like titanium alloys are the most often utilized materials.

2.2.2 Big data

Big data analytics is the process of finding patterns, trends, and linkages in huge datasets that are hidden by conventional data management methods and technologies. Big data approaches include the collection of a large dataset with potentially billions of parameters. Data are gathered from machines, sensors (temperature, vibration, humidity, etc.), controllers, and operators at all stages of production in the form of structured, unstructured, or semi-structured data for the use of big data analytics in any manufacturing process (Wu et al. 2014). The firm can monitor, trace, and reconstruct the whole supply chain process, thanks to this technology. Since it might result in losses owing to inefficiencies in manufacturing processes, the supply chain is thus seen to be more significant to the manufacturers. Big data is a crucial instrument for transforming the conventional production system into smart manufacturing systems, according to the literature. An overview study highlighting the use of big data in smart manufacturing was provided by Nagorny et al. (2017). They had the opinion that big data analysis in smart manufacturing is a vast, multidisciplinary discipline that encompasses a variety of topics. They claimed that data created at various stages of any system include significant information and may be utilized for system analysis. The foundation for decision assistance may also be provided by generated data. Data may be used for process monitoring and anomaly detection. Moreover, the researchers thought that a dataset may be used as a key performance indicator for model improvement and perhaps even for diagnosis. By using case studies in the semiconductor manufacturing industry, Moyne and Iskandar (Moyne & Iskandar 2017) have shown how big data analytics may be used for smart manufacturing. It was noted that the semiconductor manufacturing industry (SMI) uses very intricate and dynamic machinery and methods to address production quality challenges. They demonstrated how SMI is using big data approaches to enable new capabilities like predictive maintenance and enhance fault detection skills. The utilization of subject matter knowledge is also made feasible for achieving online manufacturing solutions, they claim, and the big data aspect has the ability to give high-quality solutions. They also think that advancements in big data ideas, such the digital twin, will make it possible to do analytics with subject-specific knowledge. According to McKinsey (McKinsey 2017), crucial data sets are used by any of the critical technologies, including automated in-plant logistics, supply chain management, data-driven predictive maintenance, automation, digitalized process control, smart planning and agile operations, and digital performance management, to transform any factory into a smart manufacturing facility. A research study on the use of big data analytics for smart factories was provided by Gao et al. (2020). The development of big data analytics was detailed in their study from the viewpoint of the industrial data lifecycle, commencing with data collection, transmission, management, processing, and learning. Their research also emphasized the specific issues that each stage's amount, diversity, and dependability of data presented as well as the pertinent answers that were put forward. Also, the issue of big data security is examined from both technological and governmental angles. To demonstrate the benefits of big data, five industrial case studies are presented: data for decision-making, defect prevention-based work scheduling, digital twins for production system performance, prediction of maintenance-related faults and failures, and validation of production equipment. It was suggested that the full potential of big data be further explored in light of the recognition of the significance of data in companies and academics.

According to O'Donovan et al. (2015), the data-driven revolution is causing industrial businesses to evolve into smart industries. It was found that the incorporation of cutting-edge technology like the Internet of Things (IoT) and cyber-physical systems into operational processes enabled the measurement and monitoring of the real data coming from all of the factories, producing unusual quantities of data. It makes it possible for industrial facilities to handle the demands of a remarkable rise in data output and possess the analytical methods necessary for obtaining information from these datasets. Big data, according to their suggestion, may be used to enhance a variety of manufacturing processes, including design, process planning, maintenance and diagnostics, quality management, scheduling, and even virtual manufacturing. A research on big data for intelligent industrial decision-making was provided by Li et al. (2022). An introduction to the use of big data-driven technology in intelligent manufacturing was provided. This introduction provides a framework for assessing big data-driven technology and empowers practitioners to use it to their advantage. Together with important benefits and internal motivation, they also demonstrated the potential of big data-driven technologies in the intelligent manufacturing sector. They concluded that big data analytics is one of the main technologies in artificial intelligence (AI), which enhances the market competitiveness of the manufacturing sector, and they proposed a decision analysis framework based on industrial big data-driven technology. According to their observations, this technology may assist people make wise judgments in a variety of complicated production situations by extracting hidden information and prospective talents from the industrial sector. In order to demonstrate the importance of big data in smart manufacturing, Winkler et al. (2021) conducted an industrial-academic study. Twenty-two replies were gathered from various application fields in an industry and academic survey that was performed. The replies provided underscored the need of providing fault detection and fault classification techniques based on notions of historical and present-day data analysis. The survey's findings highlighted current advantages and impending difficulties for big data applications. In the end, they came to the conclusion that big data analytics is used by several firms to gather and analyze data in order to monitor processes, optimize decisions, and lower risks and expenses in their manufacturing systems.

2.2.3 Robotic-assisted manufacturing

In conventional businesses, physical labor is used to do the majority of the job, which takes a lot of time and costs a significant amount of money. Also, the work's quality and efficiency are far from acceptable. Modern enterprises are using cutting-edge methods and technology to decrease labor hours and increase accuracy and efficiency in order to solve this issue. The adoption of innovative manufacturing techniques, including 3D printing and sophisticated robots, has recently increased across a range of industrial sectors, dramatically altering how things are produced. Robots are mostly used in industrial processes such as material handling, processing, assembly, and inspection. Robots with grippers are used in material handling to transport materials and load and unload tasks into machines. Robots use tools in processing processes to carry out tasks like welding or spray painting. Due of the high expense of using human labor in these processes, assembly and inspection are two more industrial applications where the usage of robots is anticipated to rise. Reprogramming the robots between batches allows for the production of a variety of product styles since the robots are programmable. Due to their precision and time-saving advantages, robot utilization in computer-integrated production has increased significantly in recent years. The market situation revealed that there is a constant rise in demand for output that is quick, precise, and of high quality and that contemporary automobile businesses should employ robots to achieve this expectation. Robots are capable of movement in three or more axes and are automatically controlled by programming. For usage in industrial automation applications, they may be either stationary or mobile.

A research study by Karabegovi (2016) showed the significance of industrial robots in the growth of the automobile sector. He said that China was the first nation to use industrial robot technology, which is growing exponentially every year. They proposed that the significance and variety of robotic applications in the automotive sector would eventually lead to the creation of smart automation and smart factories, which will create other high-quality goods quickly and in huge quantities in addition to automobiles. He also predicted that China will eventually utilize the most industrial robots, completely modernizing and automating manufacturing procedures in the car sector. A research study by Kulkarni et al. (2019) was conducted to show how automation has recently been used in the automobile manufacturing sector. They looked at how robotic automation is now used and also tried to come up with better ways to use automation in the future to cut down on time and human interaction. They found that robots may be used in a variety of vehicle manufacturing processes to automate the loading of wheels, monitor actual problems, tune and reorganize machines, aid in decision-making, help in robotic-aided assembly, and, most significantly, ensure industrial safety. They eventually proposed that the process has evolved to become more effective, stable, and adaptable thanks to robotic automation and artificial intelligence. For the depiction of continuous-path production jobs for robotic fillet welding, Kuss et al. (2017) introduced an innovative technique. A rule-based pattern recognition system was created for robotic-aided fillet welding, enabling automated feature detection, the creation of manufacturing data, and the derivation of geometric constraints from the product's CAD model. The transformation of geometric restrictions into basic primitives like lines, arcs, and B-spline curves makes it easier to integrate robot program planning systems from a complicated CAD environment. It is also possible to get information on welding gaps along the joint geometry. Many test scenarios for welding product assemblies were taken into account in order to promote the usability and efficacy of autonomous program planning in the robotic welding technique. Bartos et al. (2021) provide a summary of robotic uses. Their research provided a perspective on the many kinds and significance of robots in the automotive sectors. Incorporating robots into the automotive sector has been shown to boost yearly production rates, improve process accuracy, eliminate human involvement, and lower occupational dangers. The research classified robots as manipulation robots, technical robots, universal robots, and special robots. Last but not least, perspectives and advancements on the use of robots in the automobile industry are covered. This includes applications for robots in handling, machining, painting, transport, and assembly.

An investigation on the employment of robots in welding was carried out by Hong et al. (2014). Robots were described as automated, programmed equipment that could do both welding and material handling, automating the whole welding process. Silva et al. (2010) created a parallel-structure robot. The created system's axes were free and independent, so flaws in one axis won't affect the others, negating the need for pricey actuators. Further lowering the maintenance costs, any damage to one axis won't stop development in the other axes. The software created by Rubinovitz and Wysk increased the welding process's productivity and efficiency (Rubinovitz & Wysk 1988). Due to the removal of the unique point description system, their solution lowered the overall amount of time needed for online programming. Moreover, the robot programming system's setup expenses were reduced. Lee et al. (2011) created a mobile welding robot that can optimize itself depending on the available workspace. In addition, the improved design's robot was useful in building shipyard buildings and was 13% lighter than the original design.

Several research studies have shown the usage of robotic arms in AM or the 3D printing industry additive manufacturing (AM). Zhang et al.'s (2016) ABB robot was used to transport a filament-based extrusion printing head control unit. A computational framework called Robotic Studio was created to enable manufacture, modeling, and optimization. The created system was utilized to create acrylonitrile–butadiene–styrene (ABS) components, but it could also create carbon fiber composite parts. Wu et al.'s (2017) robotic arm was also utilized to create polylactic acid (PLA) items from filament without the usage of a support framework. A computational tool capable of breaking down the models into several unsupported submodels was developed using C++. The decomposition method made it possible to choose the printing direction as well. Urhal et al. (2019) have provided a review paper outlining the many uses of robots in AM. The review research described robotic-aided manufacturing in several common AM methods such as photopolymerization, direct energy deposition, and extrusion-based procedures. The research also demonstrated the usage of robot hybrid systems, which integrate additive and subtractive processes into a single operation that is carried out on the same platform.

2.2.4 Simulation

Simulation is an imitation technique that employs specialized software in the manufacturing or other business processes. Models of the genuine components or systems are created in the computer and then examined using predetermined input circumstances. The results of the analytical procedure provide the necessary data. It may be used to calculate the intended methodology's output performance and to contrast the numerous alternative solutions to find the best one for the given challenge at hand. The key goal is to comprehend how the system as a whole has changed as a result of local adjustments or changes in any one system component. It is ranked as the second most popular system in manufacturing science among managers (after real manufacturing) because it can be used to assess the impact of investments in equipment and facilities, including plant layout, warehouses, and distribution centers. Simulation is commonly employed in the industrial sector due to its many benefits. In an earlier research study, Pedgen et al. (1995) suggested that modeling, simulation, and analysis be used to examine and acquire data on the complex system. Moreover, before putting new operational or resource regulations into practice in the real world, they may be developed and tested. The fact that knowledge and insight were gained without upsetting the system itself is most significant. The goal of simulation in the product development process, according to Bodein et al. (2013), is to ensure that the generated product and component will fulfill the necessary standards. The integration of CAD and computer-aided engineering (CAE) information, as well as the general structure of the engineering procedures that call for the value chain's assistance in locating pertinent product information, was made possible by the simulation-based idea. Nyemba (2002) provided a research study demonstrating the value of modeling and simulation in supporting decision-makers in manufacturing companies. Before putting a design or operational performance study into effect, they contend that simulation is a necessary tool. They spoke about how important simulation tools are for determining the complexity of the system and forecasting its performance for system management. Moreover, they said that when compared to conventional production methods, simulation tools are also capable of offering great levels of flexibility and productivity. Four firms from the United Kingdom and the United States that have effectively employed modeling and simulation methods also provided a performance assessment of the techniques' application in their research. They also offered a perspective on how to use the methods in Zimbabwe's industrial firms. Simulation was projected to be a potent tool for the decision support system in a research study published by Fowler and Rose (2004), particularly in highly complex systems like the automotive, aerospace, electronics, and semiconductor sectors. The applicability and difficulties of simulation processes in industrial organizations were discussed by Mourtzis et al. (2014). They claim that simulation may be used in computer-aided manufacturing, the product and production life cycle, ergonomics, layout design, material flow, process simulation, and supply chain simulation. Since they enable the virtual testing and validation of various products, processes, and manufacturing system configurations, they came to the conclusion that simulation-based studies are crucial for digital manufacturing systems. The ability of the simulation tools examined in the authors' research to create more effective and adaptable systems was mentioned. A research demonstrating the significance of simulation in production systems was also presented by Mourtzis (2020). They provided a research report outlining the crucial parameters for the creation and use of simulation methods in industrial systems. Furthermore, included are the most recent technical and research methodologies in manufacturing engineering. Discussions also covered the significance of modeling and simulation in the development and management of manufacturing systems as well as the shift to digital production in the context of the Fourth Industrial Revolution. Alquraish (2022) has offered a critical study of modeling and simulation of industrial processes and systems. They looked at the tried-and-true initiatives to effectively apply modeling and simulation techniques to increase manufacturing productivity and quality, especially with the changes in contemporary factories. They also assessed the effectiveness of contemporary technologies used in manufacturing systems for enhancing design and production. The simulation community should also be aware of the difficulties and obstacles, which are also emphasized. Last but not least, the framework of the contemporary manufacturing industries also reflects the current practices, innovations, improvements, and impending possibilities.

2.2.5 Artificial intelligence

In AI, jobs that were formerly completed by manual labor are now carried out by computers. The goal of AI is to mimic human abilities like pattern recognition and decision-making. AI also offers the benefit of processing massive volumes of data that are beyond the capacity of humans. It gives computers the ability to mimic human behavior, learn from previous mistakes, and adapt to new input circumstances (Equbal et al. 2021b). Deep learning, neural networks, and natural language processing are a few typical examples of AI. AI may be used in manufacturing companies to forecast and enhance failure prediction and maintenance planning by applying it to production data. This may lead to a decrease in the price of maintaining manufacturing lines and quality control. Further advantages may be gained, such as reducing material waste and accurately predicting future demand. Since people and robots must work closely together in industrial production situations, AI and manufacturing are a natural fit. The application of AI in manufacturing is a new trend, and several associated research projects are in the works, according to a literature review. Almost 60% of industrial organizations, according to Jha et al. (2021), may be adopting AI technologies. They found that AI ensures high-quality goods and decreases industrial downtime. They also affirmed that manufacturing businesses employ AI-based solutions to increase the effectiveness of their job. Thus, they came to the conclusion that AI has a significant influence on smart manufacturing and maintenance in the production environment and also helps the manufacturers by assisting them in delivering defect-free goods on schedule. A literature study was given by Javaid et al. (2022) to demonstrate the significance of AI in Industry 4.0. They demonstrated that companies are increasingly focused on enhancing product uniformity, productivity, and operational cost reduction, and that these goals may be met by a cooperative effort between AI and humans. They made reference to how AI provides appropriate data for decision-making and alerts engineers to potential errors. They thought the study would make researchers, practitioners, students, and business people more aware of the advantages AI has for many sectors. Arinez et al. (2020) describe the existing quo and prospects for AI in advanced manufacturing. The paper's purpose was to provide three crucial bits of knowledge: (i) A systematic view was provided for data analysis and process dependencies at various levels through AI, and (ii) after reviewing the application of AI to demonstrate manufacturing problems, (iii) challenges and opportunities were identified to indicate the factors affecting AI implementation in manufacturing. The research offers in-depth analyses on a variety of subjects, including throughput and quality, guiding control in human–robotic collaboration, process diagnosis and monitoring, and, finally, advancements in materials engineering aimed at incorporating material property into process modeling and control. The use of AI in the buying, supply management, and manufacturing processes was explored by Kehayov et al. (2022). They contend that recent developments in AI and the production of vast amounts of data in the industrial sector have made it possible to integrate new tools into the supply chain system, enhancing the processes used to make items. They point to how the use of AI in manufacturing is paving the way for digitalization on plant floors, smart factories, and smart manufacturing.

2.3 DISCUSSION AND CONCLUSION

Traditional manufacturing is an industrial process that uses a lot of physical labor to turn raw materials into final goods. Also, the supply chain system and communication between the many departments are inefficient, which adds to the process' time-consuming nature. The process is also poorly defined, employs average resources, and achieves average efficiency. The amount of automation, software, and machinery used was minimal, which needlessly reduced the output's quality. Modern companies are going toward automation by altering their technology and making use of contemporary instruments in order to overcome these drawbacks and to compete in the cutthroat industry. By leveraging the technologies that incorporate computers in the production facilities, conventional factories or traditional manufacturing organizations are moving into smart factories. As a result, the manufacturing sectors adopted increased degrees of design flexibility, quick design modifications, and digitalization. It also makes it possible for contemporary manufacturing to have robust cyber security, intelligent automation, and crucial modeling and simulation. This chapter highlights the gap that currently exists between the conventional manufacturing system and the smart manufacturing system and discusses the contributions of these systems and associated technologies. The essential components and tools utilized in smart manufacturing systems are addressed, and relevant research is also provided. It is anticipated that the research study would assist researchers and professionals in determining the advantages of use in industrial settings. Google Scholar, Scopus, and the Web of Science are used to compile all of the research papers.

The use of crucial contemporary tools in the modern manufacturing industries was discussed in this study, including AI-based computer vision systems, robotic-assisted operations like robotic welding and robotic-assisted assembly of vehicle components, AM systems, modeling and simulation software, and big data technology. According to a review of the available literature on AM, AM can build any complex-shaped component in less time and with greater accuracy (Equbal et al. 2013, 2021a). According to research, AM can even be used to construct a whole mold or structure. Present-day uses of AM technology include electrodes for electrical discharge machining, investment casting, and injection molding, as stated in above AM's section. Comparing the outcomes to the conventional technologies in use, they were also exact and cost-effective (over time). A camera-integrated monitoring system, which lowers or eliminates production faults, is also made possible by AM. The manufacture of welds with different compositions was aided by the use of an ultrasonic-assisted AM process, which also allowed for the analysis of the process's microstructure and texture development (Sridharan et al. 2016). The foundation for a new advancement in the machining process was created by the preparation of EDM electrodes utilizing AM components and electroplating technology. The findings in terms of MRR, TWR, and surface quality are better, but dimensional accuracy still has to be improved, according to the literature. Big data technology allows for the examination of such systems using data produced at various stages. The foundation for decision assistance may also be provided by generated data. Also, utilizing this technology, process monitoring and anomaly detection may be accomplished. It was also shown that big data approaches are being used by the smart manufacturing sector to enhance defect detection capabilities and enable new features like predictive maintenance. It was also said that the data-driven revolution, which is made possible by big data technology, is changing the current industries into smart industries. Robotic-aided manufacturing, which is often utilized in material handling, processing activities, assembly, and inspection, improves the precision and efficiency of the task. Robots have been used to automate a variety of processes in the automobile industry, including the loading of wheels, defect monitoring in real-world settings, machine tuning and restructuring, decision-making, robotic-aided assembly, and, most importantly, industrial safety. The axes of the robots were found to be free and independent; thus, a fault in one axis would not influence the other axes, avoiding the need for expensive actuators (Silva et al. 2010). Recent research revealed that AM, or 3D printing, now uses robotic arms (Zhang et al. 2016). The technique of simulating involves creating computerized models of genuine components or systems and analyzing them under certain input circumstances. The results of the analytical procedure provide the necessary data. With modeling and simulation, the proposed methodology's output performance is estimated, and alternative solutions to find the best one for a given situation are contrasted. It demonstrated how changes in one area or in a specific system component may be used to examine changes in the whole system. It is also noted as the second most well-liked method in manufacturing science among managers (after real manufacturing). Moreover, research revealed that modeling, simulation, and analysis were done to examine and compile data on the complex system. Moreover, before putting new operational or resource regulations into practice in the real world, they may be developed and tested. It has been found that the primary goal of simulation in the product development process is to ensure that the generated product and component will fulfill the necessary standards and that any desired adjustments can only be agreed upon during the simulation phase. The introduction of AI has shown that jobs that were formerly completed by manual labor are now carried out by computers, considerably improving both efficiency and accuracy. Also, it was noted that AI has the benefit of being able to handle vast volumes of data that are beyond the scope of human intelligence. Deep learning, neural networks, and natural language processing are a few common examples of AI. AI may be used in industrial enterprises to forecast failures, enhance failure forecasting, and schedule maintenance using production data. As a result, manufacturing line maintenance expenses were reduced and quality assurance was attained. Additional advantages may be attained, such as reducing material waste and accurately estimating future demand.

2.4 CHALLENGES

Recent studies highlighted the value and relevance of smart technologies in business, which convert such businesses into smart businesses and their production systems into smart production systems. Despite the many benefits of the technologies mentioned, there are several restrictions and difficulties that have been mentioned:

1. For the use of AM in complicated industries demanding composites and light components, there are only a few materials readily accessible. Also, the cost of manufacturing is significant, particularly during setup.
2. High data storage costs, the fact that large data is unstructured, and compliance issues are the key issues with the deployment of big data technologies.
3. Affordability, the difficulty of getting the right training, and safety are significant restrictions to robotic aided production.
4. The complexity of certain software programs and the lack of background knowledge in probability and statistics among some users restrict the adoption of modeling and simulation.
5. AI can learn over time using data and previous experiences, but it is not capable of taking a novel method.

Smart tools are sophisticated and need specialized skills in production in addition to the aforementioned factors.

REFERENCES

Ablyaz, T. R. Shlykov, E. S. Muratov, K. R. Osinnikov, I. V. 2022. Study of the structure and mechanical properties after electrical discharge machining with composite electrode tools. *Materials* 15(4): 1566.

Alquraish, M. 2022. Modeling and simulation of manufacturing processes and systems: overview of tools, challenges, and future opportunities. *Engineering, Technology and Applied Science Research* 6: 9779–9786.

Ang, J. Goh, C. Saldivar, A. Li, Y. 2017. Energy-efficient through-life smart design, manufacturing and operation of ships in an industry 4.0 environment. *Energies* 10 (5): 610.

Arinez, F. J. Qing, C. Gao, X. R. Xu, C. Zhang, J. 2020. Artificial intelligence in advanced manufacturing: current status and future outlook. *Journal of Manufacturing Science and Engineering* 142: 1–16.

Bartoš, M. B. V. Bohušík, M. Stanček, J. Ivanov, V. Peter, M. 2021. An overview of robot applications in automotive industry. *Transportation Research Procedia* 55: 837–844.

Bodein, Y. Rose, B. Caillaud, E. 2013. Explicit reference modelling methodology in parametric CAD system. *Computers in Industry* 65: 136–147.

Cannizzaro, D. Varrella, A. Paradiso, S. 2021. In-situ defect detection of metal additive manufacturing: an integrated framework. *IEEE Transactions on Emerging Topics in Computing* 1: 99.

Cioffi, R. Marta, T. Piscitelli, G. Petrillo, A. Fabio, D. F. 2020. Artificial intelligence and machine learning applications in smart production: progress, trends, and directions. *Sustainability* 2: 492.

ElMaraghy, H. A. 2005. Flexible and reconfigurable manufacturing systems paradigms. *International Journal of Flexible Manufacturing Systems* 17 (4): 261–276.

Equbal, A. Akhtar, S. Equbal, A. 2020. A reflection on the use of additive manufacturing in nephrology for education and surgical planning. *Apollo Medicine* 17: 264–266.

Equbal, A. Akhter, S. Equbal, M. A. Sood, A. K. 2021a. Application of machine learning in fused deposition modeling: a review. In *Fused Deposition Modeling Based 3D Printing*, ed. H. K. Dave, J. P. Davim, 445–463. Springer Nature.

Equbal, A. Akhter, S. Sood, A. K. Equbal, I. 2021b. The usefulness of additive manufacturing (AM) in COVID-19. *Annals of 3D Printed Medicine* 2: 100013.

Equbal, A. Dixit, N. K. Sood, A. K. 2013. Rapid prototyping application in manufacturing of EDM electrode. *International Journal of Scientific & Engineering Research* 4: 1–4.

Equbal, A. Equbal, M. I. Badruddin, I. A. Algahtani, A. 2022. A critical insight into the use of FDM for production of EDM electrode. *Alexandria Engineering Journal* 61: 4057–4066.

Equbal, A. Equbal, M. I. Sood, A. K. 2019. An investigation on the feasibility of fused deposition modelling process in EDM electrode manufacturing. *CIRP Journal of Manufacturing Science and Technology* 26: 10–25.

Equbal, A. Sood, A. K. Shamim, M. 2015. Rapid tooling: a major shift in tooling practice. *Journal of Manufacturing and Industrial Engineering* 3–4: 1–9.

Fowler, J. Rose, O. 2004. Grand challenges in modelling and simulation of complex manufacturing systems. *Simulation* 80 (9): 469–476.

Gao, M. Li, L. Wang, Q. Ma, Z. Li, X. Liu, Z. 2022. Integration of additive manufacturing in casting: advances, challenges, and prospects. *International Journal of Precision Engineering and Manufacturing-Green Technology* 9: 305–322.

Gao, X. R. Wang, L. Helu, M. Teti, R. 2020. Big data analytics for smart factories of the future. *CIRP Annals—Manufacturing Technology* 69 (2): 1–25.

Hong, S. T. Ghobakhloo, M. Khaksar, W. 2014. Robotic welding technology. Comprehensive Materials Processing 6: 1–43.

Javaid, M. Haleem, A. Singh, R. P. Suman, R. 2022. Artificial intelligence applications for industry 4.0: a literature-based study. *Journal of Industrial Integration and Management* 1: 83–111.

Jha, A. K. 2021. Artificial Intelligence (AI) in manufacturing. *International Journal of Innovative Research in Engineering & Multidisciplinary Physical Sciences* 9 (3): 155–160.

Karabegović, I. 2016. The role of industrial robots in the development of automotive industry in China. *International Journal of Engineering Works* 3 (12): 92–97.

Kehayov, M. Holder, L. Kocha, V. 2022. Application of artificial intelligence technology in the manufacturing process and purchasing and supply management. *Procedia Computer Science* 200: 1209–1217.

Kennedy, G. 2010. Introduction of advanced manufacturing technology: a literature review. *Sabaragamuwa University Journal* 1: 116–134.

Kulkarni, A. A. Dhanush, P. Chetan, B. S. Thamme, G. C. S. Shrivastava, P. 2019. Recent development of automation in vehicle manufacturing industries. *International Journal of Innovative Technology and Exploring Engineering* 8: 410–413.

Kussa, A. Dietza, T. Ksensowb, K. Verlc, A. 2017. Manufacturing task description for robotic welding and automatic feature recognition on product CAD models. *Procedia CIRP* 60: 122–127.

Lee, D. Seo, T. Kim, J. 2011. Optimal design and workspace analysis of a mobile welding robot with a 3P3R serial manipulator. *Robotics and Autonomous Systems* 59: 813–826.

Lee, J. Davari, H. S. Pandhare, V. 2018. Industrial artificial intelligence for industry 4.0-based manufacturing systems. *Manufacturing Letters* 18: 20–23.

Li, C. Chen, Y. Shang, Y. 2022. A review of industrial big data for decision making in intelligent manufacturing. *Engineering Science and Technology: An International Journal* 29: 101021.

Lu, Y. Morris, K. C. Frechette, S. 2016. *Current standards landscape for smart manufacturing systems.* National Institute of Standards and Technology, NISTIR 8107: 39.

McKinsey. 2017. *Ops 4.0: manufacturing's future, made by people.*

Moldavska, A. Torgeir, W. 2017. The concept of sustainable manufacturing and its definitions: a content-analysis based literature review. *Journal of Cleaner Production* 166: 744–755.

Mourtzis, D. 2020. Simulation in the design and operation of manufacturing systems: state of the art and new trends. *International Journal of Production Research* 58 (7): 1927–1949.

Mourtzis, D. Doukas, M. Bernidaki1, D. 2014. Simulation in manufacturing: review and challenges. *Procedia CIRP* 25: 213–229.

Moyne, J. Iskandar, J. 2017. Big data analytics for smart manufacturing: case studies in semiconductor manufacturing. *Processes* 5 (2017): 39.

Nagorny, K. Monteiro, P. Barata, J. Colombo, W. A. 2017. Big data analysis in smart manufacturing: a review. *International Journal of Communications, Network and System Sciences* 10: 31–58.

Nyemba, R. W. 2002. The role of modelling and simulation in decision-making for manufacturing enterprises, techniques and challenges for the manufacturing industries. *Techniques and Challenges for the Manufacturing Industries* 11–21.

O'Donovan, P. Leahy, K. Bruton, K. O'Sullivan, T. J. 2015. Big data in manufacturing: a systematic mapping study. *Journal of Big Data* 2 (20): 1–22.

Pedgen, C. D. Shannon, R. E. Sadowski, R. P. 1995. *Introduction to simulation using SIMAN.* McGraw Hill.

Rubinovitz, J. Wysk, R. A. 1988. Task level off-line programming system for robotic arc welding-an overview. *Journal of Manufacturing Systems* 7: 293–306.

Salmi M. 2021. Additive manufacturing processes in medical applications. *Materials* 14: 191.

Silva, E. B. Filho, F. A. R. Lima, E. J. Bracarense, Q. 2010. Development of parallel robots for welding. *ABCM Symposium Series in Mechatronics* 4: 693–699.

Singh, S. Kumar, P. Singh, J. 2021. An approach to eliminate shell cracking problem in fused deposition modeling pattern based investment casting process. *IOP Conference Series: Materials Science and Engineering* 1091: 012035.

Sridharan, N. Wolcott, P. Dapino, M. Babu, S. S. 2016. Microstructure and texture evolution in aluminum and commercially pure titanium dissimilar welds fabricated using ultrasonic additive manufacturing. *Scripta Materialia* 117: 1–5.

Urhal, P. Weightman, A. Diver, C. Bartolo, P. 2019. Robot assisted additive manufacturing: a review. *Robotics and Computer Integrated Manufacturing* 59: 335–345.

Winkler, D. Korobeinykov, A. Novák, P. Lüder, A. Biffl, S. 2021. Big data needs and challenges in smart manufacturing: an industry-academia survey. *IEEE Xplore*, 21466915.

Wu, C. Dai, C. Fang, G. Liu, Y. J. Wang, C. 2017. RoboFDM: a robotic system for support-free fabrication using FDM. *IEEE International Conference on Robotics Automation* 1175–1180.

Wu, X. Zhu, X. Wu, G.Q. Ding, W. 2014. Data mining with big data. *IEEE Transactions on Knowledge and Data Engineering* 26: 97–107.

Zhang, G. Q. Spaak, A. Martinez, C. Lasko, D. T. Zhang, B. Fuhlbrigge T. A. 2016. Robotic additive manufacturing process simulation-towards design and analysis with building parameter in consideration. *IEEE International Conference on Automation Science Engineering* 16467358.

Chapter 3

A review of sustainable approach to smart manufacturing systems

Pankaj Kumar Tyagi, Sheeba Jilani, Faisal Talib, and Mohan Kumar

3.1 INTRODUCTION

In order to optimize strategies, produce new goods, reduce development times, increase resource efficiency, or provide more tailored products, businesses and governments are being encouraged to embrace and develop new technologies as Industry 4.0 emerges (Xu 2022; Khin and Hung Kee 2022). An important factor contributing to this trend is the move from product-oriented businesses to service-oriented ones, along with increased usage of the Internet of Things (IoT) (Jamshed et al. 2022; Koohang et al. 2022; Jahromi et al. 2023). In addition, more and more automated, networked, and intelligent methods are being used rather than computer-assisted procedures. A smart manufacturing (SM) system is enabled by the IoT, cyber-physical systems (CPS), cloud-based solutions, advanced robotics, the Internet of Services (IoS), big data analytics, augmented reality, 3D printing, virtual reality, and artificial intelligence (AI) (Kumari et al. 2022; Sahoo and Lo 2022). Many fields have begun referring to smartness as well as smart systems as a result of the use of these terms. In the scientific literature, the term "smart" has been used to describe a wide variety of goods, including smartphones, televisions, watches, and other electronic items. Figure 3.1 shows the schematic layout of the interconnection of SM systems in Industry 4.0.

These concepts often refer to advancements made in a particular field because of the creation, application, and use of new technologies such as ICT, intelligent gadgets, AI, and revolutionary industrial and managerial concepts (Singh et al. 2023). In general, SM refers to processes that are fully integrated, cooperative, and able to adapt instantly to changing market conditions, supply chains, and consumer demands. A new generation of manufacturing sectors is using more intelligent and smart technologies; SM systems have been found to increase productivity in manufacturing by 17–20% by maximizing the machine's usage and optimizing the energy used (Tuptuk and Hailes 2018). These systems include smart warehouses, intelligent machines, and digitally developed production facilities. According to this paradigm, cyber-physical production systems (CPPS) are SM systems. The emerging global trend of SM includes intelligent manufacturing in

DOI: 10.1201/9781003495314-3

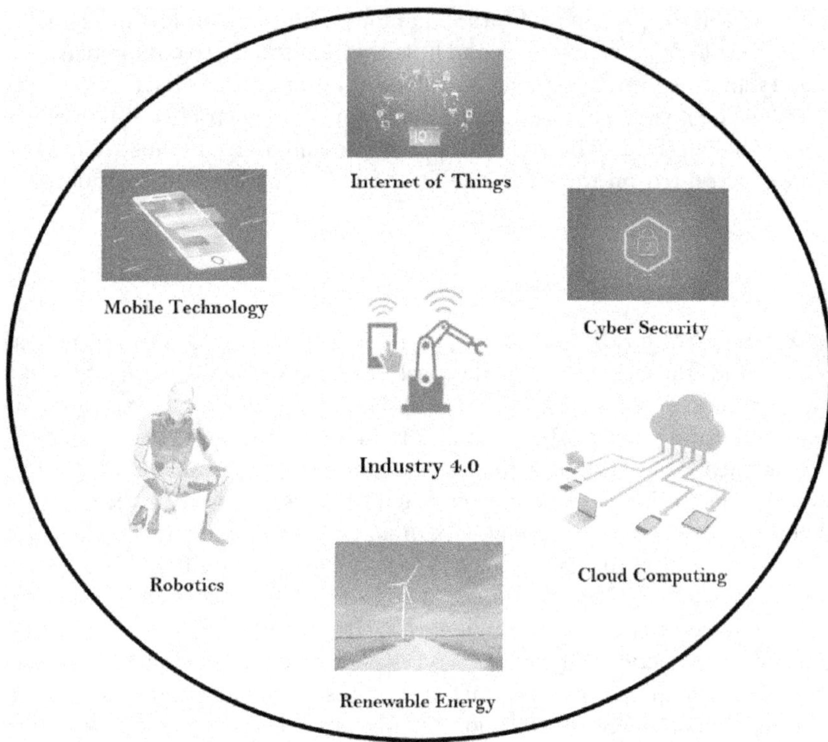

Figure 3.1 Smart Manufacturing Systems.

China, smart factories in Japan and Korea, industrial Internet of Things, and SM in the United States (Kim 2022). When properly applied, SM goes beyond what the world preaches now because it is more accessible, inexpensive, assessable, inventive, frictionless, and collaborative (Shukla and Shankar 2022). The incorporation of technologies powered by smart devices into industry remains a significant and multifaceted problem. Due to a lack of capacity and knowledge, specifically small and medium-sized firms (SME) frequently suffer issues. The evaluation and adoption of smart devices that are appropriate for the intended use case are vital stages in such initiatives. Here, a methodical description of smart connected devices assists the selection process by categorizing the devices and assisting businesses in understanding smart devices' precise needs before mass acquisition. A requirement list may be created by matching the characteristics of these devices with a range of potential properties.

Environmental concerns still pose hurdles and obstacles to the use of sustainable production management procedures (Caiado and Quelhas 2020; Kaushal and Chakrabarti 2022). An overview of SM and its current implementation challenges is provided in this chapter, along with an analysis of the gap between the present and future manufacturing systems and a discussion of the technologies related to it and how they affect SM. To comprehensively understand the rapidly expanding technology and cover all its dimensions, a study of the most recent advancements in this field, their effects, implementation challenges, and future directions is presented. A bibliometric analysis is also presented on the sustainability of SM to understand the published work in this domain.

3.1.1 Literature survey

Sustainable SM has been receiving more and more attention from the business world and academics in the past few years. The concept of "smart manufacturing" incorporates powerful data analytics with seamless, intelligent monitoring, diagnostics, and autonomous operations in order to enable omnipresence, intelligent diagnosis, and autonomous production (Qu et al. 2019). There have been a number of significant innovations in recent years, including zero inventory production, digital twin-based manufacturing, augmented reality, virtual metrology (VM), and blockchain-based smart contracts (Li et al. 2022; Yang et al. 2022). In order to achieve SM, one of the most important enabling technologies is virtual metrology (VM). Through the use of prediction algorithms, VM is used to directly estimate attributes that affect the quality of a product after it has been manufactured. The VM system establishes a virtuous loop through its virtual–physical link, which continually updates data so that the system can be real-time; in addition, it produces useful insights that can be applied in the physical production environment. The combination of sustainable SM and digital twin can pose a challenge. Using information management systems (IMSs) as the foundation for sustainable SM, Yang et al. (2022) propose a digital twin strategy focused on the use of big data in energy-intensive industries (EIIs) (Ma et al. 2022). Leng and their colleagues suggest a solution for open architecture flow-type smart manufacturing systems (SMS) called digital twins-based remote semi-physical commissioning (DT-RSPC) (Leng, Wang, et al. 2021; Leng, Zhou, et al. 2021). In order to enable remote semi-physical commissioning, a digital twin system is created. A case study of a digital twins-based remote semi-physical commissioning of a smartphone manufacturing line serves to verify the suggested technique. The outcomes show that commissioning an innovative flow SMS is much more sustainably accomplished by integrating the open architectural design paradigm with the suggested digital twins-based strategy. Big data applications are often focused on data-related technologies, such as cloud computing, data purification, and data mining, whereas digital twins rely on network-physical integration technologies like simulation, virtual reality, augmented reality, and CPS (Chen et al. 2014).

Even if SMS can handle many of the issues and complexities faced by the present industries, there are still some challenges with their use. Therefore, a bibliometric analysis is presented here to understand the sustainability of SM systems. Bibliometric techniques are now well-established scientific specialties and form a crucial component of research evaluation methodology, particularly in the scientific and applied sectors (Ellegaard and Wallin 2015). Scholars conduct a bibliometric analysis to identify the knowledge base in a specific topic, to examine its research front, and to map out and produce networking structures of the scientific community interested in the topic (Aria and Cuccurullo 2017). In view of the aforementioned factors, a bibliometric study was undertaken on SM systems and their sustainability to understand the trends and patterns of research in this subject.

3.2 RESEARCH METHODOLOGY

The research approach employed here comprises two related steps. It first describes how the papers were chosen as the method for data collection and then how these publications were analyzed.

3.2.1 Data collection approach

Most often, data gathering begins with choosing the database that will hold the input data. As in the works of earlier researchers (Baz and Iddik 2022; Fahimnia et al. 2015; Tseng et al. 2019), our sources are the bibliographic records from the Scopus database, which has been deemed trustworthy in these earlier works. Additionally, Scopus is the largest database of scientific peer-review literature, hosting more than 22,000 titles and high-impact scientific studies from international publishers (Elsevier.Com).

Right search phrases were chosen to query the database. We searched the title and abstract fields of the publications using various keyword combinations, some of which are listed in Table 3.1. Scopus is a huge database that contains thousands of journals; therefore, filtering the results is necessary after data collection. Only papers written in English with some significance in the titles were chosen in the selected list and exported in the Bibtex file format.

Following the scanning of the 802 articles returned by the above search results and the elimination of duplicates and articles written in languages other than English, 431 articles were ultimately chosen for this study.

Table 3.1 Searched Outcomes

Searched keywords	*No. of papers*
(TITLE-ABS-KEY ("smart manufacturing") AND TITLE-ABS-KEY (sust*))	448
(TITLE ("smart manufacturing") AND TITLE-ABS-KEY (sust*))	135
(TITLE ("smart manufacturing") AND TITLE (sust*))	54
(TITLE-ABS-KEY ("smart manufacturing") AND TITLE (sust*))	165

3.2.2 Data analysis

Following the researcher (Ellegaard and Wallin 2015) the bibliometric analysis is made. The Bibliometrics and Scientometrics software enable comprehensive quantitative investigation in these fields (Aria and Cuccurullo 2017). The software tools utilized in our research are freely available and can be used by anybody to identify current trends and gaps in the databases and publications that are already available. The benefit of clustering publications utilizing open access tools like Mendeley and VOSviewer is that it doesn't require highly developed computer skills or an extensive understanding of clustering algorithms (Baz and Iddik 2022).

3.3 RESULTS OF BIBLIOMETRIC ANALYSIS

The results of the bibliometric analysis are presented in this section after being divided into distinct subsections. The most prevalent words used in titles and keywords, top journals, top organizations, and a thorough co-citation analysis were all covered in this descriptive study.

3.3.1 Descriptive analysis

Discussing writers, their affiliations, nations, and the sources of the papers, as well as the documents, their contents, and bibliographies, is the focus of this descriptive analysis. A thorough review of the entire discipline is the main goal of this effort. In order to do this, we had to include all document types (Table 3.2) that were extracted using a keyword search on the entire page and may provide a number of significant scientific contributions.

After 2017, researchers' interest in the sustainability of SMS increased significantly. Prior to that, yearly publications in this field remained in single digit, but starting in 2017, the number of publications increased dramatically (Table 3.3). This indicates that the field is still being explored, as there were likely no publications in the searched area prior to 2008.

Table 3.2 Typology of Published Selected Work

Typology	*Number of documents*	*%*
Journal articles	217	50
Conference papers	154	36
Reviews	30	07
Book chapters	21	05
Books	3	01
Editorial/letters/short surveys, etc.	6	01
Total	431	100

Table 3.3 Scientific Evolution over the Years in the Searched Domain

Year	*Number of documents*
2008	2
2011	1
2012	2
2013	2
2014	4
2015	8
2016	6
2017	23
2018	34
2019	44
2020	83
2021	97
2022	113

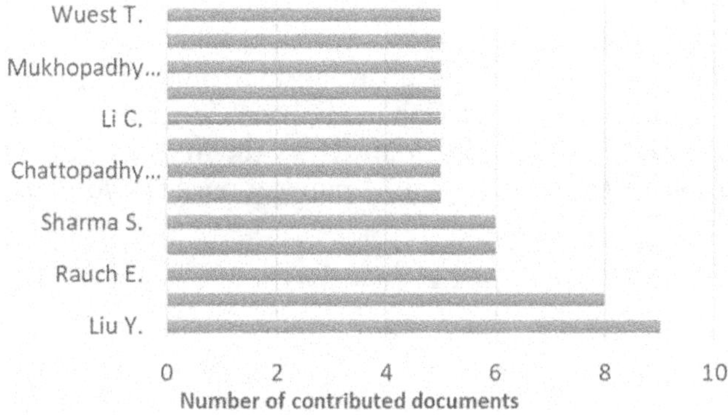

Figure 3.2 Most Contributing Authors.

3.3.2 Influence of researchers

The 431 analyzed articles were written by a total of 1,297 researchers, which is almost triple the number of published articles. Thirteen authors have more than five publications. Liu Y. is the most contributing author, with nine publications, followed by Zhang Y. and Ren S., with eight publications and six publications, respectively (Figure 3.2). Zhang Y. is the most cited author with 496 citations, followed by Ren S. with 457 citations from his six contributed documents.

3.3.3 Affiliation analysis

Based on an affiliation study, it can be seen that the researchers of the aforementioned documents are affiliated with 935 different organizations, of which three have made more than five contributions. With eight articles, the Faculty of Operation and Economics of Transport and Communications, Department of Economics, University of Zilina, Zilina, Slovakia, contributed the most. The School of Mechanical and Automotive Engineering, Qingdao University of Technology, Qingdao, China, and The Cognitive Labor Institute, New York City, NY, United States, each contributed five articles, but the Department of Management and Engineering, Linköping University, Linköping, Sweden, received the highest citation of 301 among the contributing organizations with its three contributions only.

Also these researchers belong to 68 countries, and 16 countries have more than 10 publications, while 7 have more than 20 publications each. The United States is the most contributing country, with 97 articles, followed by India and China, with 61 and 53 documents, respectively (Figure 3.3).

3.3.4 Sources relevancy

The selected 431 documents were published in 207 sources, out of which 15 sources have published five or more articles. 35 documents out of the 431 selected documents got published in "Sustainability (Switzerland)." IFIP Advances in Information and Communication Technology ranked second with a publication of 16 articles, followed by the *Journal of Self-governance and Management Economics* with 14 publications. The *Journal of Cleaner*

Figure 3.3 Most Contributing Countries in the Researched Domain.

Production seems to be the most promising resource, with 1,094 citations and 11 publications out of the selected documents.

3.3.5 Highly cited articles

The analysis of the 431 selected documents shows that 43 of these documents have 50 or more citations (Figure 3.4), while 19 articles have 100 or more citations (Table 3.4). The document written by Kusiak (2018) with the title "Smart Manufacturing" is the most cited document with 623 citations, followed by the researcher Kamble et al. (2018) with the title "Sustainable Industry 4.0 Framework: A Systematic Literature Review Identifying the Current Trends and Future Perspectives" with 578 citations, and the article written by Lopes de Sousa Jabbour et al. (2018) with the title "Industry 4.0 and the Circular Economy: A Proposed Research Agenda and Original Roadmap for Sustainable Operations" with 490 citations.

The article written by the researcher Ghobakhloo (2020) seems more promising with 220 TC/year in the selected domain.

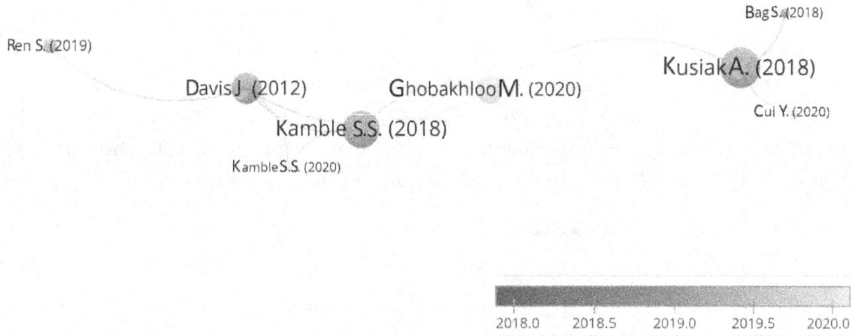

Figure 3.4 Citation Clustering of the Most Cited Documents.

Table 3.4 Top Most Cited Documents in the Domain

Paper	*Source*	*Total citations*	*TC/ year*
(Kusiak 2018)	*International Journal of Production Research*	623	155
(Kamble et al. 2018)	*Process Safety and Environmental Protection*	578	144
(Lopes de Sousa Jabbour et al. 2018)	*Annals of Operations Research*	490	122
(Davis et al. 2012)	*Computers and Chemical Engineering*	485	48
(Ghobakhloo 2020)	*Journal of Cleaner Production*	440	220

3.3.6 Statistics of keywords

The keyword usage statistics provide information on the major terms used in the texts. The researchers employed a total of 1,191 keywords in these papers, of which 17 were used in ten or more (Figure 3.5) and 7 in 20 or more. With 187 occurrences, "Smart Manufacturing" is the most frequently occurring keyword, followed by "industry 4.0," "sustainability," "sustainable manufacturing," "internet of things," "artificial intelligence," and "big data," which occurred in the order 115, 63, 44, 34, 26, and 26, respectively.

3.3.7 Intellect inputs: co-citation analysis

The intellect input is provided below in three different groups based on the most-cited sources, authors, and articles among the 431 papers that were collected. Six references out of the 20,723 were mentioned more than nine times in the gathered papers. The article entitled "Opportunities of Sustainable Manufacturing in Industry 4.0," published in the source *Procedia CIRP* in the year 2016, is the most referred, with 21 referrals among these 431 papers. In addition, the 431 articles that we gathered cited 8,570 sources, of which 35 were cited in more than 50 of the articles we gathered, and 12 were cited in more than 100 of the articles we chose. Our chosen documents cited the *Journal of Cleaner Production* the most, receiving 773 citations, followed by the *International Journal of Production Research and Procedia CIRP*, which received 572 and 413 citations, respectively. Additionally, these 431 papers referenced 30,290 writers, of which 18 were cited in over 100 documents (Figure 3.6). In these chosen articles, Liu. Y. received

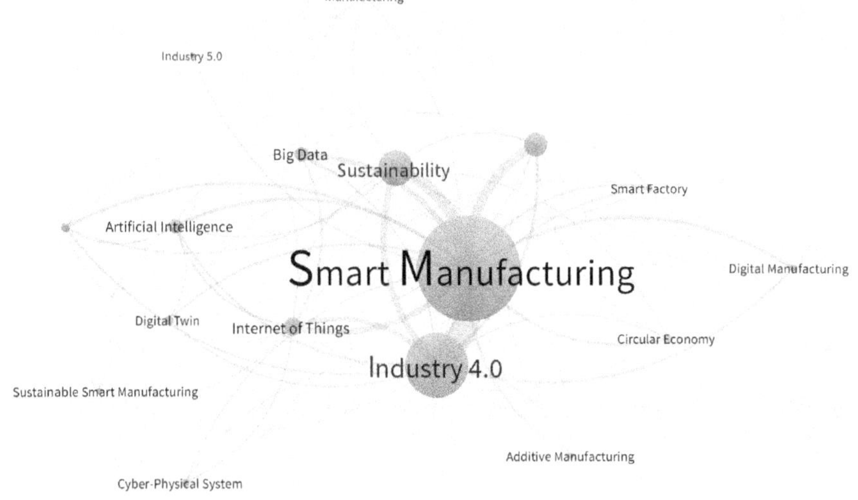

Figure 3.5 High-Occurrence Keywords.

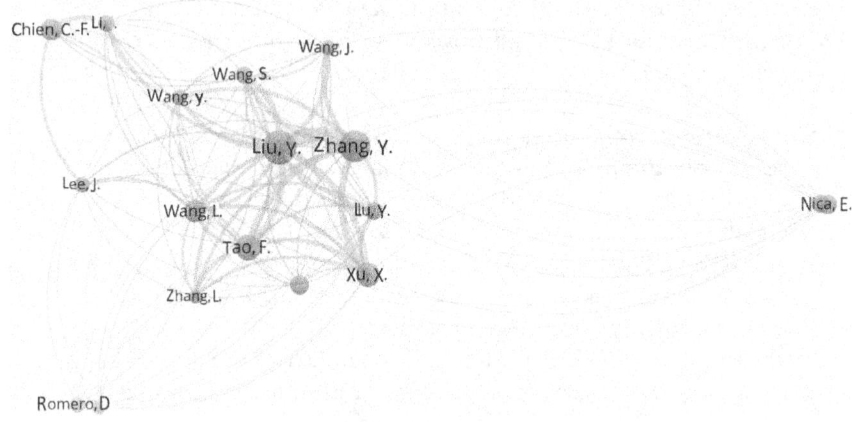

Figure 3.6 Co-citation Analysis.

245 referrals, followed by Zhang. Y. and Tao. F., who received 224 and 184 referrals, respectively.

The above bibliometric analysis highlights the conceptual structure of the selected field and its evolution over a period of time. It also indicates that the field is still being explored, as there were likely no publications in the area of sustainability of SM systems prior to 2008.

3.4 CHALLENGES RELATED TO SM SYSTEMS

Though SM systems still encounter some difficulties, they are capable of handling many of the difficulties and complexities faced by current industries (Oztemel and Gursev 2020; Cohen et al. 2019). Several dependent variables have been used to determine whether there will be security concerns, a lack of system integration, a low return on investment in new technology, and financial issues when building new SMS or retrofitting existing industries with SM technology. SM systems are a desirable target for cyber-attacks since they include important technology and sensitive information (Maggi et al. 2021; Ren et al. 2017). An attacker can control the whole production facility once they gain access to the network because there is no segmentation of the network.

This groundbreaking innovation's technological viability hinges on the smart device itself and its production method. However, examining the applicability of this strategy across a wide range of procedures for a range of products reveals significant opportunities in the SMS's latter stages. It is possible to identify the value proposition for a particular smart product and SMS when combined with an assessment of the financial advantages. It is essential for smart manufacturing and manufacturing data analytics to have accurate, high-quality, and large-volume data; yet this data is often collected and analyzed "externally," omitting details about the transformation process (Wan et al. 2018). A few of the challenges in SM are discussed below:

Human–robot collaboration: It is possible to interact physically and securely with humans in the workplace with the help of robots called "lightweight robots" and foster new paradigms for human–machine collaboration (Haesevoets et al. 2021). During the implementation of any CPS system or industrial robotic system, priority should be given to reducing any mechanical, electrical, thermal, noise, vibration, radiation, material, or substance hazards.

Security issues in SM: An integrated manufacturing system that shares information between manufacturing or machining units and with consumers is called a SMS. The setup uses the internet for this purpose and is specifically designed for this reason. Information sharing via the internet requires end-to-end encryption of data and information at many points in the system. A unique identity must be maintained at every point, and a worldwide unique ID must be established. It is therefore crucial to make sure that every node of the network is protected from outside threats and data exploitation. Creating networked systems, like SMS, requires top-priority security for all components.

System integration: An obstacle to the installation of a SMS is the integration of the latest technological equipment with existing ones. As a result of the compatibility between old and brand-new equipment, several issues arise when using SM technologies. There is a possibility that modern gadgets use a different protocol than outdated gear. In addition, a stronger communication system is needed due to machine-to-machine communication and system interconnectivity. IPv6 connection is necessary for most contemporary production systems in order to enable devices to be linked concurrently.

Interoperability: The capacity of several systems to comprehend and utilize one another's features independently is known as interoperability (Zeid et al. 2019). With the help of this capability, they may interchange data and information without regard to who made their hardware or software. I4.0 interoperability is divided into four levels: operational, systematical, technological, and semantic (Rezaei et al. 2014). Operational interoperability is related to I4.0's conceptual framework and has to do with CPS. A systematic approach to interoperability involves standards, directives, key approaches, and models. Technical interoperability clearly specifies platforms and resources for technical, ICT, and associated software contexts. Using semantic interoperability, people and institutions can share information across levels (de Mello et al. 2022). Communication protocols and standards must be synchronized in order to achieve effective interoperability. In order for a system to be interoperable, it must have different transmission bandwidths, operational frequencies, modes of communication, hardware capabilities, etc.

Researchers Shukla and Shankar (2022) mentioned various critical success factors for SM system implementation:

Real-time operating system: A quick and responsive software element called a real-time operating system arranges jobs and manages resources efficiently. Such a technological element supports real-time applications within predetermined time limits and ensures the existence of SM systems.

Cyber security: Information technology and electronic information security are both terms for cyber security. A key component of an SM system is a network of networked devices where data and information are constantly exchanged and evaluated. Having strong cyber security will give you the confidence to make changes.

Execution of interoperability in the system: The capacity to effortlessly transmit technical and commercial information across the production system is known as interoperability. Additionally, it makes it possible for manufacturing partners to properly and speedily share their equipment, helping the financial element of the SM system deployment.

Optimization of resources: Utilizing resources efficiently involves choosing the finest possibilities. Optimization can reduce expenses, increase stability, and maintain the system's functionality, all of which are crucial for SMEs.

Resource sharing: When companies with similar interests pool their resources, this is referred to as resource sharing. This can make it more affordable for SMEs to access advanced resources.

Smart modular workstation: A smart modular workstation is composed of various modules, each of which uses technology and smart tools to do specific tasks on its own. In this way, SMEs can adopt SM systems where they are most needed without having to undergo a total technological transformation.

Data management capability: Real data management entails having quick access to data as needed. A discipline for data management should be established and developed inside an organization. This is essential because the SM system depends on data to function.

Mature IT system: An IT system that has reached maturity is one that is completely functional, serves the intended function, and performs with no problems. The development of IT systems has a big impact on the efficient exchange of information and the ability to operate in real time, opening up opportunities for SM systems to be integrated into existing systems.

Digitization: The development of connected platforms and devices has made it possible for manufacturing to continue to digitize. Digitization increases output and improves process visualization, enabling processes to move more quickly and meet deadlines that are more demanding.

Smart product, service, and supply chain: Any system or service is said to be "smart" when it has some degree of autonomy and interactive capability. Such systems and services are necessary for an organization's output to be flexible, automated, and self-regulating.

Align the SM initiative with organization strategy: Any technology change's ability to succeed is largely dependent on the company's predetermined goal and vision. The implementation process will be sped up by bringing the organizational strategies of the company and the notion of SM into alignment.

Right use of financial capabilities: The ability to make sound financial decisions requires a combination of attitude, abilities, knowledge, and efficiency. SMEs encounter some limitations when it comes to investment and appropriate use; adopting SM systems can aid further in growing their annual turnover.

Agile and experienced decision-makers: Decision-makers that are agile can act quickly and decisively in complicated situations while refining and adjusting as needed by the organization. These persons encourage the deployment of the SM system with their reliability and assistance.

Continuous top management commitment: It alludes to managerial conviction and ongoing support for innovation and participation in an organization's SM system.

Continuous quality improvement: Companies adopt it as a management philosophy to cut down on waste, boost productivity, and improve employee and customer happiness. A potential justification for integrating the SM system into the current one is the improvement in product quality, which is a fundamental component of the SM system.

Mass customization of goods and services: A marketing strategy known as "mass customization" refers to a company's ability to quickly generate items that are tailored to each individual customer's needs and preferences, offering them the option to choose features from a basic package. Only the SM system allows for the execution of this technique to be envisioned.

Proper division of employees under training: Financial resources can be saved by having a well-established training system as per staff deployment. Due to the limited investment capacity of SMEs, these strategies are essential for achieving organizational excellence.

Active knowledge management system: Transferring and codifying tacit knowledge of new and current processes is challenging. Having access to educated professionals and their instruction is essential for such system change because this can only be accomplished via experience.

Supporting innovation capabilities and workload management: Giving employees the tools they need to use their creativity to develop and implement fresh ideas might help a company's technological revolution. A balanced workload will encourage independent thought and concern for the organization.

Reliable partners with knowledge and experience: While adopting the SM system in an organization with a shared interest, a business partner's ability to be trusted and reliable can generate useful counsel and support.

Government support and regulations: These control businesses through laws and regulations, thus governing how they can conduct themselves and make investments, and require businesses to abide by these rules. For the purpose of facilitating the adoption of such technology, the government can enact new regulations and offer assistance.

Technology support infrastructure: This kind of infrastructure consists of the tangible technological resources, services, and supplies required for a company to run smoothly. It is more likely to be possible to implement complicated systems, such as SM systems, with this architecture.

Accounting corporate social responsibility: Through an excellent perspective on environmental factors, SM practices lay the groundwork for advancements in business sustainability. The results of the SM system include minimal waste, reused raw materials, and no surplus production.

Personnel social and motivational support: Humans do not always welcome and embrace change. Consequently, educating people about the benefits of technology advancements and treating them as integral components of this transformation help streamline the SM adoption process.

Consumer readiness: A consumer may want to buy a specific product. Consumers are increasingly focused on smart products and services in this technology age, which can encourage businesses to use SM systems to meet consumer needs.

Sustainability consideration: Implementing the SM system is a radical transformation that industries are pursuing to endure the competitive environment. Sustainability research examines how economic, social, and environmental factors affect a business, thus making SM an ongoing, holistic endeavor.

Implementation strategy: Organizations have long-term and short-term objectives. To follow stated strategies that result in competitive advantages and improved performance, an organization should establish and arrange its organizational structure, resources, and working culture. An effective strategic plan will accelerate the process of SM system implementation.

Project activity model: With the purpose of streamlining the process of implementing a new system, these models typically enlist the assistance of consultants.

Change management: Structural shifts occur both vertically and horizontally. Managing this radical shift is essential because it examines how the change will affect people and processes and connects it with the established plan for SM system integration.

Enterprise performance matrix: An organization's activity, capability, and overall quality are represented by numbers and data in a performance matrix. An organization must review its performance and decide on its main area of concentration because doing so will make it easier to determine whether system change is necessary and how successful the organization is.

Decentralization: One facility, huge premises, and significant investment are characteristics of centralized production. Yet decentralized manufacturing structures are more advantageous since they allow for the division of the production process among several sites and the use of smart workstations in accordance with the needs and preferences of the organization. Decentralization is essential for the development of intelligent, autonomous units.

Among the above critical success factors for SM implementation, researchers Shukla and Shankar (2022) found that "government support and regulation" and "continuous top management support" are the two most substantial and dynamic factors for employing SM system.

3.5 OPPORTUNITIES AND FUTURE DIRECTIONS

The potential market for SM systems is enormous, and they are being extensively used in the global production system. Modern industrial systems have grown to be large live robotic ecosystems, with little to no human presence in certain instances. Resources with smart capabilities can do the same and alert the system in real time if there are problems on a production line. The system can get continuous feedback on the state of smart goods by continuously monitoring them. In order to accomplish shared objectives, the various parts of the system communicate and interact with one another. Moreover, SMS allows more information and data to be extracted from the system than ever before, allowing information and data to be turned into knowledge. In modern times, most constraints are related not to hardware or connections between entities but to the development of SMS. They are extremely complex, and there may be limitations on the process and memory power. Cyber-physical systems have enabled all types of shopfloor entities to be virtualized and connected to each other, making it simple for monitoring and controlling to be accomplished. The connecting of many different databases and how they ought to work together to provide improved and more effective performances is one of the primary issues. Therefore, it is essential to take things a step further and focus more energy on modeling, optimizing, and standardizing production systems to make these systems more sustainable.

3.6 CONCLUSION

A rapid change in manufacturing systems has been caused by the convergence of information technology, industrial production technology, and human creativity. The use of innovative and cutting-edge technology to research sustainable SM in industries has already been extensively documented in the literature. Technology such as big data and other cutting-edge technologies are frequently used in SM to achieve sustainability. When digital twin technology is used to replicate a physical shopfloor, real-time data cannot be collected without taking into consideration how the data is collected and generated. It is continually necessary for manufacturing systems to make use of the latest techniques to boost responsiveness and agility in order to fulfil the variety of needs in a market that is constantly changing cost-effectively; also, we have to be more concerned about the sustainability of these systems as "We do not inherit the Earth from our ancestors; we borrow it from our children."

3.7 DECLARATION OF COMPETING INTEREST

The authors affirm that they have no known financial or interpersonal conflicts that would have seemed to have an impact on the research presented in this study.

ACKNOWLEDGMENTS

P.K.T., S.J., and M.K. are thankful to the editors of the book who continuously motivated and supported this work.

REFERENCES

Aria, Massimo, and Corrado Cuccurullo. 2017. "Bibliometrix: An R-Tool for Comprehensive Science Mapping Analysis." *Journal of Informetrics* 11 (4): 959–75. https://doi.org/10.1016/j.joi.2017.08.007.

Baz, Jamal El, and Sadia Iddik. 2022. "Green Supply Chain Management and Organizational Culture: A Bibliometric Analysis Based on Scopus Data (2001–2020)." *International Journal of Organizational Analysis* 30 (1): 156–79. https://doi.org/10.1108/IJOA-07-2020-2307.

Caiado, Rodrigo Goyannes Gusmão, and Osvaldo Luiz Gonçalves Quelhas. 2020. "Factories for the Future: Toward Sustainable Smart Manufacturing." In *Responsible Consumption and Production. Encyclopedia of the UN Sustainable Development Goals*, edited by W. Leal Filho, A. M. Azul, L. Brandli, P. G. Özuyar, T. Wall, 239–250. Springer. https://doi.org/10.1007/978-3-319-95726-5_108.

Chen, Min, Shiwen Mao, Yin Zhang, and Victor CM Leung. 2014. *Big Data [Electronic Resource]: Related Technologies, Challenges and Future Prospects*. Springer EBooks.

Cohen, Yuval, Maurizio Faccio, Francesco Pilati, and Xifan Yao. 2019. "Design and Management of Digital Manufacturing and Assembly Systems in the Industry 4.0 Era." *International Journal of Advanced Manufacturing Technology* 105: 3565–77. https://doi.org/10.1007/s00170-019-04595-0.

Davis, Jim, Thomas Edgar, James Porter, John Bernaden, and Michael Sarli. 2012. "Smart Manufacturing, Manufacturing Intelligence and Demand-Dynamic Performance." *Computers and Chemical Engineering* 47. https://doi.org/10.1016/j.compchemeng.2012.06.037.

de Sousa Jabbour, Lopes, Ana Beatriz, Charbel Jose Chiappetta Jabbour, Moacir Godinho Filho, and David Roubaud. 2018. "Industry 4.0 and the Circular Economy: A Proposed Research Agenda and Original Roadmap for Sustainable Operations." *Annals of Operations Research* 270 (1–2). https://doi.org/10.1007/s10479-018-2772-8.

Ellegaard, Ole, and Johan A Wallin. 2015. "The Bibliometric Analysis of Scholarly Production: How Great Is the Impact?" *Scientometrics* 105 (3): 1809–31. https://doi.org/10.1007/s11192-015-1645-z.

Fahimnia, Behnam, Joseph Sarkis, and Hoda Davarzani. 2015. "Green Supply Chain Management: A Review and Bibliometric Analysis." *International Journal of Production Economics* 162: 101–14. https://doi.org/10.1016/j.ijpe.2015.01.003.

Ghobakhloo, Morteza. 2020. "Industry 4.0, Digitization, and Opportunities for Sustainability." *Journal of Cleaner Production* 252: 119869. https://doi.org/10.1016/j.jclepro.2019.119869.

Haesevoets, Tessa, David De Cremer, Kim Dierckx, and Alain Van Hiel. 2021. "Human-Machine Collaboration in Managerial Decision Making." *Computers in Human Behavior* 119. https://doi.org/10.1016/j.chb.2021.106730.

Jahromi, Amir Namavar, Hadis Karimipour, and Ali Dehghantanha. 2023. "An Ensemble Deep Federated Learning Cyber-Threat Hunting Model for Industrial Internet of Things." *Computer Communications* 198 (January): 108–16. https://doi.org/10.1016/J.COMCOM.2022.11.009.

Jamshed, Muhammad Ali, Kamran Ali, Qammer H Abbasi, Muhammad Ali Imran, and Masood Ur-Rehman. 2022. "Challenges, Applications, and Future of Wireless Sensors in Internet of Things: A Review." *IEEE Sensors Journal* 22 (6): 5482–94. https://doi.org/10.1109/JSEN.2022.3148128.

Kamble, Sachin S, Angappa Gunasekaran, and Shradha A Gawankar. 2018. "Sustainable Industry 4.0 Framework: A Systematic Literature Review Identifying the Current Trends and Future Perspectives." *Process Safety and Environmental Protection* 117. https://doi.org/10.1016/j.psep.2018.05.009.

Kaushal, Ishaan, and Amaresh Chakrabarti. 2022. "System Modelling for Collecting Life Cycle Inventory (LCI) Data in MSMEs Using a Conceptual Model for Smart Manufacturing Systems (SMSs)." *International Journal of Precision Engineering and Manufacturing-Green Technology* 10: 819–34. https://doi.org/10.1007/s40684-022-00489-x.

Khin, Sabai, and Daisy Mui Hung Kee. 2022. "Identifying the Driving and Moderating Factors of Malaysian SMEs' Readiness for Industry 4.0." *International Journal of Computer Integrated Manufacturing* 35 (7). https://doi.org/10.1080/0951192X.2022.2025619.

Kim, Hyunjung. 2022. "Performance from Building Smart Factories of Small- and Medium-Sized Enterprises: The Moderating Effects of Product Complexity and Company Size." *International Journal of Operations & Production Management* 42 (10): 1497–520. https://doi.org/10.1108/IJOPM-10-2021-0654.

Koohang, Alex, Carol Springer Sargent, Jeretta Horn Nord, and Joanna Paliszkiewicz. 2022. "Internet of Things (IoT): From Awareness to Continued Use." *International Journal of Information Management* 62. https://doi.org/10.1016/j.ijinfomgt.2021.102442.

Kumari, Meet et al. 2022. "The Cyber Physical Systems and Industrial Internet of Things (IIoT): A Revolution." In *Security and Resilience of Cyber Physical Systems*, edited by Sajal Bhatia Krishan Kumar, Sunny Behal, Abhinav Bhandari, 1st Edition, 1–12. Chapman and Hall/CRC. https://doi.org/10.1201/9781003185543.

Kusiak, Andrew. 2018. "Smart Manufacturing." *International Journal of Production Research* 56 (1–2): 508–17. https://doi.org/10.1080/00207543.2017.1351644.

Leng, Jiewu, Dewen Wang, Weiming Shen, Xinyu Li, Qiang Liu, and Xin Chen. 2021. "Digital Twins-Based Smart Manufacturing System Design in Industry 4.0: A Review." *Journal of Manufacturing Systems* 60: 119–37. https://doi.org/10.1016/j.jmsy.2021.05.011.

Leng, Jiewu, Man Zhou, Yuxuan Xiao, Hu Zhang, Qiang Liu, Weiming Shen, Qianyi Su, and Longzhang Li. 2021. "Digital Twins-Based Remote Semi-Physical Commissioning of Flow-Type Smart Manufacturing Systems." *Journal of Cleaner Production* 306. https://doi.org/10.1016/j.jclepro.2021.127278.

Li, Lianhui, Bingbing Lei, and Chunlei Mao. 2022. "Digital Twin in Smart Manufacturing." *Journal of Industrial Information Integration* 26. https://doi.org/10.1016/j.jii.2021.100289.

Ma, Shuaiyin, Wei Ding, Yang Liu, Shan Ren, and Haidong Yang. 2022. "Digital Twin and Big Data-Driven Sustainable Smart Manufacturing Based on Information Management Systems for Energy-Intensive Industries." *Applied Energy* 326 (September): 119986. https://doi.org/10.1016/j.apenergy.2022.119986.

Maggi, Federico, Marco Balduzzi, Rainer Vosseler, Martin Rösler, Walter Quadrini, Giacomo Tavola, Marcello Pogliani, Davide Quarta, and Stefano Zanero. 2021. "Smart Factory Security: A Case Study on a Modular Smart Manufacturing System." *Procedia Computer Science* 180 (2019): 666–75. https://doi.org/10.1016/j.procs.2021.01.289.

Mello, Blanda Helena de, Sandro José Rigo, Cristiano André da Costa, Rodrigo da Rosa Righi, Bruna Donida, Marta Rosecler Bez, and Luana Carina Schunke. 2022. "Semantic Interoperability in Health Records Standards: A Systematic Literature Review." *Health and Technology* 12 (2): 255–72. https://doi.org/10.1007/s12553-022-00639-w.

Oztemel, Ercan, and Samet Gursev. 2020. "Literature Review of Industry 4.0 and Related Technologies." *Journal of Intelligent Manufacturing* 31 (4): 127–82. https://doi.org/10.1007/s10845-018-1433-8.

Qu, Yuanju, Xinguo Ming, Yanrong Ni, Xiuzhen Li, Zhiwen Liu, Xianyu Zhang, and Liuyue Xie. 2019. "An Integrated Framework of Enterprise Information Systems in Smart Manufacturing System via Business Process Reengineering." *Proceedings of the Institution of Mechanical Engineers, Part B: Journal of Engineering Manufacture* 233 (11). https://doi.org/10.1177/0954405418816846.

Ren, Anqi, Dazhong Wu, Wenhui Zhang, Janis Terpenny, and Peng Liu. 2017. *Cyber Security in Smart Manufacturing: Survey and Challenges*. 67th Annual Conference and Expo of the Institute of Industrial Engineers 2017.

Rezaei, Reza, Thiam Kian Chiew, Sai Peck Lee, and Zeinab Shams Aliee. 2014. "Interoperability Evaluation Models: A Systematic Review." *Computers in Industry* 65 (1): 1–23. https://doi.org/10.1016/j.compind.2013.09.001.

Sahoo, Snehasis, and Cheng Yao Lo. 2022. "Smart Manufacturing Powered by Recent Technological Advancements: A Review." *Journal of Manufacturing Systems* 64 (July): 236–50. https://doi.org/10.1016/J.JMSY.2022.06.008.

Shukla, Monica, and Ravi Shankar. 2022. "An Extended Technology-Organization-Environment Framework to Investigate Smart Manufacturing System Implementation in Small and Medium Enterprises." *Computers and Industrial Engineering* 163 (March 2021): 107865. https://doi.org/10.1016/j.cie.2021.107865.

Singh, Amrinder, Geetika Madaan, Swapna Hr, and Anuj Kumar. 2023. "Smart Manufacturing Systems: A Futuristics Roadmap towards Application of Industry 4.0 Technologies." *International Journal of Computer Integrated Manufacturing* 36 (3): 411–28. https://doi.org/10.1080/0951192X.2022.2090607.

Tseng, Ming Lang, Md Shamimul Islam, Noorliza Karia, Firdaus Ahmad Fauzi, and Samina Afrin. 2019. "A Literature Review on Green Supply Chain Management: Trends and Future Challenges." *Resources, Conservation and Recycling* 141 (September 2018): 145–62. https://doi.org/10.1016/j.resconrec.2018.10.009.

Tuptuk, Nilufer, and Stephen Hailes. 2018. "Security of Smart Manufacturing Systems." *Journal of Manufacturing Systems* 47. https://doi.org/10.1016/j.jmsy.2018.04.007.

Wan, Jiafu, Jun Yang, Zhongren Wang, and Qingsong Hua. 2018. "Artificial Intelligence for Cloud-Assisted Smart Factory." *IEEE Access* 6. https://doi.org/10.1109/ACCESS.2018.2871724.

Xu, Li Da. 2022. "Emerging Enabling Technologies for Industry 4.0 and Beyond." *Information Systems Frontiers*. https://doi.org/10.1007/s10796-021-10213-w.

Yang, Jing, Xiaoxing Wang, and Yandong Zhao. 2022. "Parallel Manufacturing for Industrial Metaverses: A New Paradigm in Smart Manufacturing." *IEEE/CAA Journal of Automatica Sinica* 9 (12): 2063–70. https://doi.org/10.1109/JAS.2022.106097.

Zeid, Abe, Sarvesh Sundaram, Mohsen Moghaddam, Sagar Kamarthi, and Tucker Marion. 2019. "Interoperability in Smart Manufacturing: Research Challenges." *Machines* 7 (2): 21. https://doi.org/10.3390/machines7020021.

Chapter 4

The status of smart agriculture

A method to enhancing management

Muhammad Farhan, Nortoji A. Khujamshukurov, Zainab Fatima, Tufail Ahmad, Muhammad Zaid, Kuzmin Anton, M. Masroor A. Khan, Mohammad Azeem, Mohammad Azad Alam, and Mohd Usman Khan

4.1 INTRODUCTION

The food crisis and population growth are the two major obstacles to sustainable development on a global scale. Water is regarded as a crucial resource for all living things because it is practically impossible for life to exist on this planet without it. Every living thing consumes food in accordance with its needs, as we all know. Due to the significance of water in our daily lives, it is essential that water resources be used effectively and that no water be wasted. Water is used by almost all industries, although the agriculture industry is the biggest water user. This industry is responsible for providing around 40% of the world's food needs, and it uses roughly 70% of the water resources on Earth for irrigation. Just as there is an increase in the global population, so is the demand for food. Farmers need to utilize more water to cultivate more crops, especially orchards, in order to fulfil the rising population's increased demand for food (Farhan et al., 2022).

Since the history of humanity is preserved, agriculture, farming, or husbandry is a crucial profession. The term "agriculture" refers to all things that were connected in a linear fashion to the human food chain. Humans are the most intelligent species alive today, and as such, their intelligence constantly spurs them to develop and change. The growth of languages, the development of the wheel, better living circumstances and lifestyles, as well as numerous other innovations were all the result of this motivation (Liu et al., 2011). Technology is used in many different ways in smart agriculture. With the help of these technologies, traditional farming methods are being replaced with efficient, dependable, and sustainable smart agriculture. A few examples are vehicle guiding, feeding animals facilities, traceability systems, food packing and inspection, remote surveillance, security control, context-aware farming, water irrigation, pesticide control, and environmental monitoring. Robotics is also being implemented in accordance with smart technologies to create more space for technological improvements in agriculture. Smart agriculture benefits from the Internet of Things (IoT), a study area that is now hot (Doitsidis et al., 2017). The agricultural reforms essentially consist of three primary areas, as shown in Figure 4.1.

DOI: 10.1201/9781003495314-4

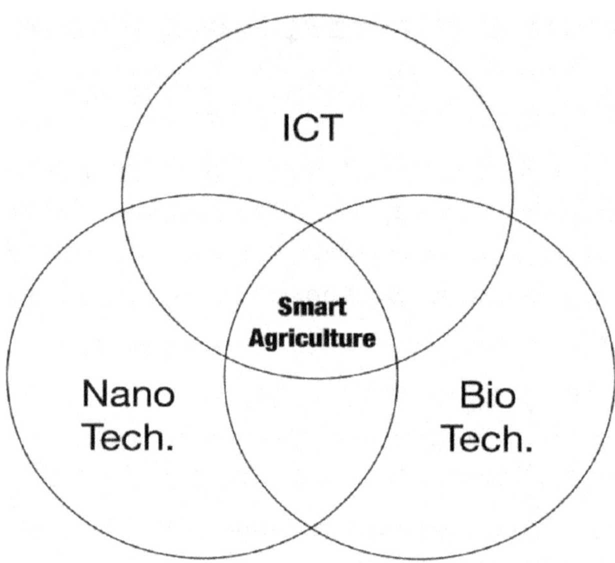

Figure 4.1 Standing of Smart Agriculture.

4.2 THE CONCEPT OF SMART

In general, we regard a computer to be smart if it does a task that we believe an intelligent person is capable of performing. Any system, process, or domain is deemed intelligent if it adheres to the six levels of intelligence.

- Adapting: The changes would be seen as environmental in the framework of smart agriculture. The phrase "adaptation" describes a modification made to meet any particular requirements.
- Sensing: The capacity to detect environmental changes or to notice any changes.
- Inferring: This term essentially refers to drawing a conclusion from data and observances.
- Learning: Learning can be used to improve prior methods after getting findings and results. It involves several kinds of information.
- Anticipating: Imagining something novel and inventive that is about to happen or, as we can say, the next stage of anything is what anticipation is all about.
- Self-Organization: Any intelligent system that has the capacity to perceive, monitor, and then adjust its parameters in response to demands is said to be self-organizing.

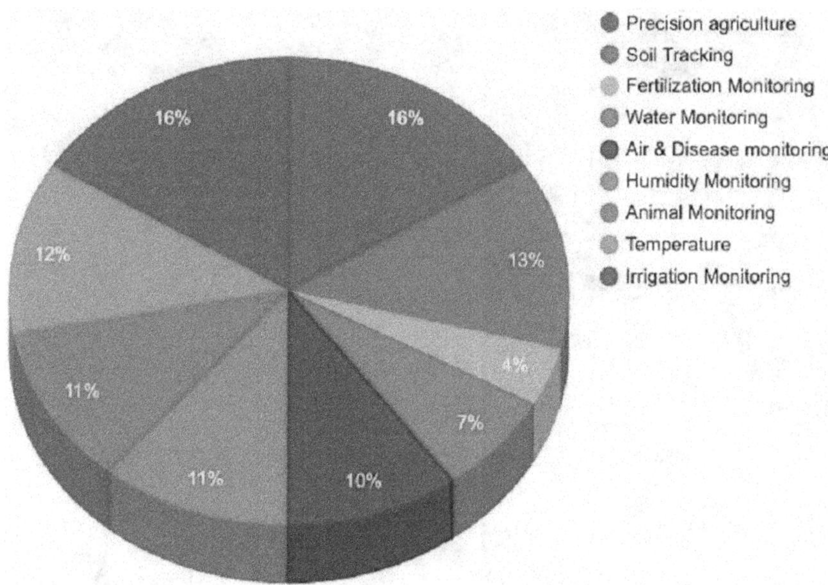

Figure 4.2 Smart Agriculture Distribution of Three Tier.

4.3 IOT IN SMART AGRICULTURE

The provision of high-speed internet, cutting-edge smartphone technology, and satellites is essential for the development of smart systems (Islam et al., 2010; Liu and Hong Peng, 2013). Researchers James et al. (2019) successfully used IoT in actual time to monitor and identify leaf diseases that impede crop growth; they achieved this through numerous satellite images or sensors placed in farms (paddy and bananas crops). Their system helped in the analysis of information and decision-making. They then sent the decisions back to the power source farmers via the internet server. The Organization of Food and Agricultural of the UN (FAO, 2017) said that a lack of adequate crop monitoring results in the loss of 20–40% of crops each year to pests, diseases, and other factors. Thus, the use of sensor and intelligent systems permits the monitoring of weather variables, fertility condition, and also the determination of the precise amount of fertilizers required for crop growth (IAASTD, 2008). The fertility of the soil is negatively impacted by the overuse of fertilizers. Farooq et al. (2020) looked at 67 research articles on the usage of IoT in various agricultural applications that have been published during 2006 and 2019. Precision agriculture accounted for roughly 16% of the studies, followed by papers on irrigation monitoring, soil tracking, 16%, 12%, 11%, and 5% on temperature, animals, humidity, and air

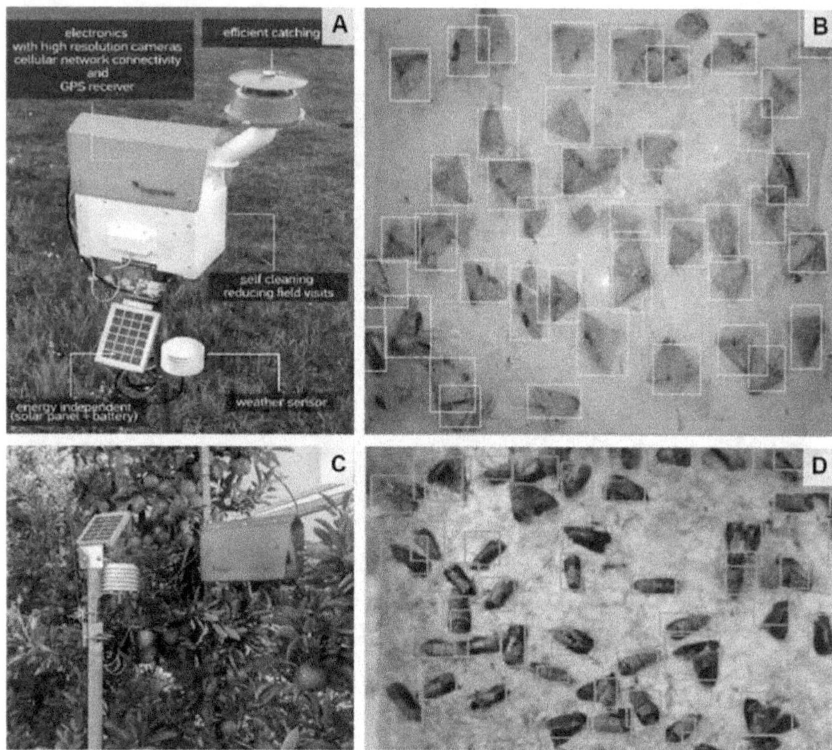

Figure 4.3 Applications of IoT in Various Areas.

and disease monitoring, and water monitoring, which made up 7% of the publications (Cox, 2002). Last but not least, only 4% of the studies included fertilization monitoring (Figure 4.3).

4.4 5G NETWORK IN SMART FARMING

Information and communication technologies have advanced significantly over the past several years in tandem with the development of smart systems (Esfahani and Asadiyeh, 2009). In the last 10 years, 3G, 4G, and NB-IoT wireless network technologies have been employed to connect smart gadgets through the IoT and exchange information for accurate assessment in the agriculture industry. However, as information has improved in both quantity and quality, due to poorer data transfer than before, the 4G network's efficacy has decreased. Generation five (5G) network represents the progression of communication networks by offering extremely fast data

Figure 4.4 Application of IoT in Smart Agriculture.

transformation speeds. When compared to other networks, the 5G network's data transmission speed is faster since its download and upload rates are nearly 100 times faster than those of the 4G network. In comparison to 6 minutes on 4G, downloading a 2-hour video would require just over four seconds on 5G. Additionally, the downlink rate in a 5G network is 20 Gbps, with a 10 Gbps characteristic (Anand et al., 2020). Utilizing a 5G network for smart applications has numerous benefits, including the following (Figure 4.4):

- Fast data transfer speeds and little latency.
- An exceptionally high connection density in relation to another network.
- Enhanced efficiency in the spectrum.
- A seamless communication performance.

4.5 CLIMATE-CONSCIOUS FARMING

A technique known as "climate-smart agriculture" (CSA) attempts to solve the three challenges of sustainable agriculture, climate change, and food security. In addition to lowering greenhouse gas emissions and boosting resistance to the consequences of global warming, the CSA seeks to raise agricultural output and incomes. Since the Food and Agriculture Organisation (FAO) initially presented the idea of CSA in 2010, it has been well known. The strategy is centered around three key pillars:

- Boosting incomes and productivity
- Improving adaptability and resistance to climate change
- Lowering greenhouse gas emissions.

One of the key components of CSA is the use of climate-resilient crops and sustainable agricultural practices. These practices include conservation agriculture, agroforestry, and integrated pest management. Additionally, CSA encourages the use of climate-smart technologies, such as precision agriculture, weather forecasting, and digital soil mapping, to increase efficiency and reduce environmental impacts (Rassati et al., 2016). Several studies have

Table 4.1 Climate-Smart Agriculture Practices and Their Reported Benefits

Sl. no.	*Pillars*	*Components*	*Implementation*	*Benefits*
1	Increasing productivity and incomes	Use of climate-resilient crops and sustainable agricultural practices	Conservation agriculture, agroforestry, integrated pest management	Increased yields, improved soil health, and reduced greenhouse gas emissions
2	Enhancing resilience and adaptation to climate change	Use of climate-smart technologies	Precision agriculture, weather forecasting, digital soil mapping	Increased efficiency and reduced environmental impacts
3	Reducing greenhouse gas emissions	Adoption of CSA practices	Sustainable agriculture, reducing greenhouse gas emissions	Environmental benefits and reduced vulnerability to climate-related shocks
4	Social and economic benefits	Adoption of CSA practices	Sustainable agriculture, reducing greenhouse gas emissions	Increased incomes, improved food security, and reduced vulnerability to climate-related shocks

shown that CSA practices can lead to increased produces, enhanced soil health and decreased emissions of greenhouse gases. For instance, an Ethiopian research discovered that the utilization of conservation agriculture practices increased maize yields by 73% while also reducing soil erosion and improving soil fertility (Alemu et al., 2020).

In addition to the environmental and productivity benefits, CSA can also have social and economic benefits for smallholder farmers. Adoption of CSA practices raised earnings, enhanced food security, and decreased sensitivity to shocks connected to climate change, according to a Kenyan research (Makokha et al., 2019). In general, community-supported agriculture is a significant strategy for attaining sustainable agriculture and tackling the issues of food security and climate change. Governments, international organizations, and private sector actors can play a critical role in promoting and supporting CSA practices and technologies to ensure a sustainable future for agriculture.

4.6 SUSTAINABLE AGRICULTURE PRACTICES IN SMART AGRICULTURE

Sustainable agriculture practices are vital for the success of smart agriculture, which utilizes advanced technologies to optimize resource use and enhance productivity while minimizing the environmental impact. Recent research has highlighted the potential of several sustainable agriculture practices to be integrated into smart agriculture systems to improve long-term productivity, profitability, and environmental sustainability.

One promising approach is the integration of precision agriculture with sustainable agricultural practices. Precision agriculture uses advanced technologies such as drones, remote sensing, and GPS to collect data on soil, weather, and plant growth (Wang et al., 2006). By integrating sustainable agricultural practices such as cover cropping and crop rotation, farmers can reduce input costs, optimize resource use, and improve soil health. For example, according to a recent research of Liu et al. (2021), integrating precision irrigation with cover cropping and crop rotation resulted in increased water-use efficiency and improved soil health, leading to higher crop yields and profitability.

Another approach is the integration of agroforestry into smart agriculture systems. Agroforestry involves the integration of trees, crops, and livestock on the same land, promoting biodiversity, soil health, and multiple sources of income for farmers. Recent research has shown that agroforestry can also reduce greenhouse gas emissions and promote carbon sequestration (Amjath-Babu et al., 2021). By integrating agroforestry with smart agriculture technologies, such as precision irrigation and nutrient management, farmers can further optimize resource use and improve long-term sustainability.

Table 4.2 Comparison of Sustainable Agriculture Practices for Smart Agriculture

Sl. no.	*Sustainable agriculture practices*	*Benefits*	*Implementation*	*References*
1	Precision Agriculture	Improved Resource Use Efficiency	Use of Drones and Remote Sensing Technologies for Precision Irrigation and Nutrient Management	Liu et al., 2021;
2	Agroforestry	Biodiversity Promotion and Carbon Sequestration	Integration of Trees, Crops, and Livestock on Same Land	Amjath-Babu et al., 2021
3	Regenerative Agriculture	Soil Health Restoration and Greenhouse Gas Reduction	Reduced Tillage, Diversified Crop Rotations, and Holistic Grazing Management	Gomiero et al., 2020
4	Cover Cropping	Improved Soil Health and Reduced Soil Erosion	Planting Cover Crops during Fallow Periods	Bashir et al., 2021
5	Integrated Pest Management	Reduced Chemical Use and Increased Biodiversity	Biological Control and Crop Rotation	Kogan and Jepson, 2007

In addition, several emerging sustainable agriculture practices, such as regenerative agriculture, are gaining attention in the context of smart agriculture (Gutierrez and Robbes, 2013). Regenerative agriculture emphasizes the restoration of soil health, biodiversity, and ecosystem services through the use of practices such as reduced tillage, diversified crop rotations, and holistic grazing management. Recent research has shown that regenerative agriculture may decrease greenhouse gas emissions, strengthen soil health, and farm profitability (Gomiero et al., 2020). By integrating regenerative agriculture with smart agriculture technologies, such as with the use of big data analytics and machine learning, farmers can maximize resource utilization and enhance long-term sustainability.

The integration of sustainable agriculture practices with smart agriculture is crucial for promoting long-term productivity, profitability, and environmental sustainability. Emerging research highlights the potential of precision agriculture, agroforestry, and regenerative agriculture as promising approaches for the integration of sustainable agriculture practices into smart agriculture systems.

4.7 ROLE OF GIS IN SMART AGRICULTURE

As discussed in [11], researchers think that geographic information system (GIS) is crucial in developing global agricultural policy. GIS and associated technologies are used to monitor crop yield, and from these monitors, farmers can obtain, for instance, maps every 10 meters. The researchers also go over the use of GIS for agriculture for agronomic, logistical, and marketing purposes. Along with other spheres of life, agriculture benefits greatly from GIS. In agriculture, GIS is used for a variety of tasks, including policy-making, erosion of soil mapping, making choices, forecasting, modeling climate change, for evaluating trimming patterns, pest management, nitrogen management, with prior supervision to reduce test errors, and generate information gathering and its evaluation, among other things (Wark et al., 2007; Yukikazu et al., 2012). GIS solutions help farmers manage their geographic information for farming, from gathering it via mobile devices to analyzing background-sensing data at their workplace by improving development, decreasing cultivating expenses, more effective handling of risk factors and handling of resources, creating higher revenue, greater savings, and prompt decision-making, among other things.

4.8 FUTURE OF AGRICULTURE: GEOGRAPHICAL AGRICULTURE INFORMATION SYSTEM (GAIS)

Nanotechnology in farming is Opara (2004) used in spatial geographic agricultural information systems (GAIS), which combines computer technology and GIS applications to provide information and boost agriculture's overall productivity. The term "GAIS" in this context refers to a collection of already-available GIS tools and a group of agricultural apps that enable agriculture users to fully make use of GAIS's analytical and predictive capabilities (Jiang et al., 2013). This includes lowering and saving input farming costs, managing resources wisely, and increasing profit and productivity. GAIS is nothing more than a package that combines a GIS tool and tools and a collection of apps specifically designed for use in agriculture (Figure 4.5). To support GAIS, developers must choose a previous GIS product or toolset and produce specialist agricultural apps. More smart and more advanced agriculture is what GAIS is expected to grow into, where spatial farming data is intelligently gathered using smart gadgets like cell phones, tablets, and PDAs, processed by intelligent GAIS tools and delivered in smart forms as well as formats like graphs with conventional tabular representations. In the future, agricultural machinery will be able to connect to GPS systems and be controlled by mobile gadgets like tablets using GAIS. The researcher's main goal in this regard will be to create GAIS smartphone apps.

4.9 NEW APPROACHES FOR PEST MANAGEMENT IN AGRICULTURE

Regardless of the agricultural style used, crop losses from pests and diseases are the main source of concern for farmers. Plant diseases, pest insects,

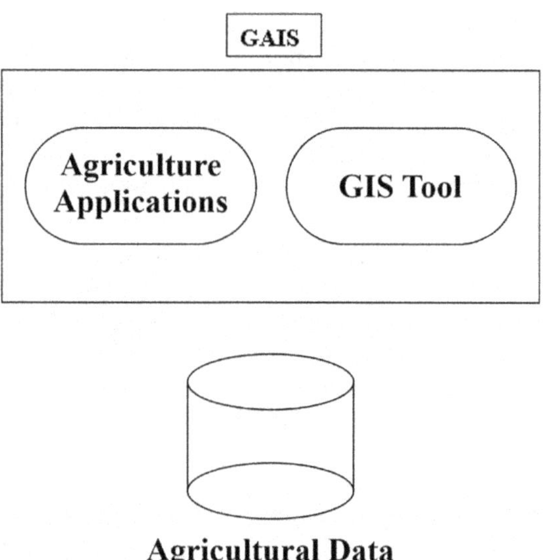

Figure 4.5 A Sample Geographic Agriculture Information System (GAIS).

and weed infestations destroy more than between 30% and 40% of the potential food production each year. This loss occurs even with the application of about 3 million pesticide tonnes each year, along with a variety of nonchemical controls like crop rotation and biological controls (Agarwal and Verma, 2020).

There are several approaches, tools, and techniques to fight and minimize pre- and post-harvest losses. With advancements in technology, tools and storage methods have changed according. Agarwal and Verma (2020) state that crop protection refers to methods, supplies, and techniques that farmers employ to safeguard their produce against pests, illnesses, and weeds (Guarnieri et al., 2011). Farmers all over the world make decisions every day about the best ways to protect their crops from weeds, which deprive them of sunlight, nutrients, and water. These decisions concern the use of diverse techniques such as microbiological pesticides, biological control, insect behavior, genetic engineering, and plant immunization of pest populations. Internet of Things (IoT), remote sensing (RS), an array of PA technologies, including GIS and artificial intelligence (AI), enable farmers to oversee pest populations. The Mississippi River Delta was mapped to identify the stress that aphid species generate, and remote sensing was utilized to anticipate the most likely areas where wheat crops might be attacked by insects (Yang et al., 2009).

Many types of insect pests can be automatically recognized and managed, claims Cardim Ferreira Lima et al. (2020). To enhance precision agriculture and integrated pest control (IPM), a number of systems have been developed. For several significant pests, automatic detection traps have been created. Exploring automatic detection, we present the many systems that are available, examples of applications, and contemporary advances, like Internet of Things and machine learning, with an emphasis on approaches for identifying pests using acoustic, infrared, and image-based sensors.

4.10 TECHNIQUE FOR STERILE INSECTS

A target pest is mass-reared and sterilized using radiation as part of the environmentally friendly sterile insect technique. The sterile males are then released by air over predetermined areas in a systematic manner, permitting them to pair up with wild females, whereupon they would not bear fruit and the number of pests would decrease. The sterility insect approach, or SIT for short, is one of the most effective insect pest management techniques ever developed. Sterilization by radiation, such as gamma or X-rays, renders mass-reared insects incapable of procreating even while they are still sexually competitive. Genetic engineering, or transgenic, procedures are not used in SIT.

4.11 MATING DISRUPTION TECHNIQUE

Female insects release sex pheromone to attract male partners for mating. The sex pheromone release by females is unique from species to species. These sex pheromones are synthetically produced for the management of insects. Each species has a unique combination of molecules that make up its sex pheromones. Typically, the female releases a sex pheromone moth, and the male locates the female by following the pheromone trail, also known as the plume. When creating "lures" (for pheromone traps), synthetic pheromones are produced, which are usually mixtures of primary and secondary chemical components meant to resemble the properties of naturally occurring pheromones. The higher the reaction from seeking males, the more closely the mix resembles that created by a pheromone-producing female, sometimes referred to as a "calling" female. In order to confuse males and hinder their capacity to find calling females, mating disruption (MD) technology employs enormous volumes of synthetically generated chemicals; nonetheless, the mixes are often limited to key components emitted by females.

4.12 MANAGEMENT OF INSECT PESTS WITH EMERGING TECHNOLOGIES

Insect monitoring is the most difficult and challenging task. Breeding patterns of insects and their high adaptability to the changing climate are driving forces for insect population. Many times, prediction phonological models to forecast insect population fail and vary from species to species (Holguin et al., 2010). Numerous publications attempt to enhance insect pest monitoring by the incorporation of electrical devices into monitoring traps. In the 1980s, optical instruments were incorporated into sex-pheromone baited traps to automatically count the number of cabbage looper Trichoplusia ni Hübner and tobacco budworm Heliothids virescent Fabricius, including those (both belonging to the Noctuidae family of insects) that were caught (Hendricks, 1985). This is the oldest-known occurrence of this process. An important advancement was made, according to Hendricks (1990), when recorded data was remotely sent to a computer. An example of this was the automated identification of the boll weevil, *Anthonomus grandis Boheman* (*Coleoptera: Curculionidae*). The first use of cameras for monitoring insect pests was made by Kondo et al. (1994) for the tobacco cutworm *Spodoptera litura Fabricius* (*Lepidoptera: Noctuidae*) and the Asiatic rice borer *Chilo suppressalis Walker* (*Lepidoptera: Crambidae*). Thanks to the tremendous advancements in technology over the past 25 years, along with the creation of camera traps for wildlife monitoring, Michele Preti et al. (2021) reported that camera devices have been used more and more for insect pest monitoring. A basic literature search was performed in a scientific database of titles published in the last century; the terms "camera" or "video" or "smart or autonomous or automatic or electrical or long distance or remote" and "tracking or detection" or "insect or pest" were the key words used. Out of less than hundred articles that were published, over 90% of the publications that were identified were published within the previous 25 years, indicating it's a relatively fresh subject. The current review focuses on insect pest observation, with camera-equipped traps in order to illustrate the advantages and disadvantages of using a camera and, thus, insects to capture photos and replace the conventional human-powered onsite trap check used in farming and forest frameworks (Ünlü et al., 2019; Shaked et al., 2018).

4.13 CONCLUSION

Smart approaches in agricultural sciences, employed in sowing up to harvesting, storing, and preserving the products, are playing leading roles in "Farm to Fork" and fulfilling the demands of the world population. It is time to extend smart technologies from industries to producing lands and farmers to train them to adopt better production methods. Traditional approaches are subjective and labor-intensive and prohibit large-scale production. Green vegetables and leaf vegetables are in more demand in the market in comparison to grain foods. The market costs of these vegetables are high due to low production, unseasonal demands, and a lack of smart approaches to production and preservation. Due to the high demand for green vegetables, farmers are harvesting unripe vegetables, i.e. potato, chili, carrot peas, and radish, which become dry and unshaped and shrink, sometimes leading to fungal infections. The sellers generally use a chemical that retains the freshness of these vegetables for a limited time. For example, unripe potatoes usually turn green and produce chlorophyll due to the presence of solanin, which behaves like a poison.

ACKNOWLEDGMENT

The facilities to conduct this study and the technical help for authoring this chapter were provided by Universiti Putra Malaysia (UPM) and Aligarh Muslim University, India, for which the authors are grateful. The advice provided throughout the chapter by Prof. Sapuan Salit (UPM), Malaysia, is much appreciated by the authors. The authors also acknowledge Dr. Mohammad Azeem of Universiti Technologi Petronas, Malaysia, for his counsel and insightful conversations.

REFERENCES

Agarwal M, Verma A (2020) Modern technologies for pest control: a review. https://doi.org/10.5772/intechopen.93556

Alemu W, Wondimu T, Tsegaye D (2020) Conservation agriculture and its contribution to climate-smart agriculture: evidence from Ethiopia. *Sustainability* 12(10): 4292.

Amjath-Babu TS, Krupnik TJ, Kaechele H, Aravindakshan S, Arunagiri S, Bhattarai BK, Zougmore RB, et al. (2021). Smart technologies for sustainable food systems. *Nature Food* 2(3): 166–172.

Anand A, Gustavo de Veciana G, Shakkottai S (2020) Joint scheduling of URLLC and eMBB traffic in 5G wireless networks. *IEEE/ACM Transactions on Networking* 28(2): 477–490.

Bashir MK, Ali S, Khalid S, Mubeen M, Ahmad S, Ahmad R (2021) Sustainable agriculture practices and their potential role in food security: a review. *Sustainability* 13(2): 604.

Cox S (2002) Information technology: the global key to precision agriculture and sustainability. *Computers and Electronics in Agriculture* 36: 93–111.

Doitsidis L, Fouskitakis GN, Varikou KN, Rigakis II, Chatzichristofis SA, Papafilippaki AK, Birouraki AE (2017) Remote monitoring of the Bactrocera oleae (Gmelin) (Diptera: Tephritidae) population using an automated McPhail trap. *Computers and Electronics in Agriculture* 137: 69–78. https://doi.org/10.1016/j.compag.2017.03.014

Esfahani L, Asadiyeh Z (2009) *The Role of Information and Communication Technology in Agriculture*. The 1st International Conference on Information Science and Engineering (ICISE 2009).

FAO (2017) *Food and Agriculture Organization of the United Nations*. https://www.fao.org/family-farming/detail/en/c/1245425/ [Assessed on: December 2, 2023]

Farhan M, Mastura MT, Ansari SP, Muaz M, Azeem M, Sapuan SM (2022) Advanced potential hybrid biocomposites in aerospace applications: a comprehensive review. In *Advanced Composites in Aerospace Engineering Applications*. Springer Nature. https://doi.org/10.1007/978-3-030-88192-4_6

Farooq MS, Riaz S, Abid A, Umer T, Zikria YB (2020) Role of IoT technology in agriculture: a systematic literature review. *Electronics* 9(2): 319. https://doi.org/10.3390/electronics9020319

Gomiero T, Pimentel D, Paoletti MG (2020). Soil health, ecosystem services and food security: critical review and implications for sustainable agriculture. *Sustainability* 12(17): 7062.

Guarnieri A, Maini S, Molari G, Rondelli V (2011) Automatic trap for moth detection in integrated pest management. *Bull Insectology* 64(2): 247–251.

Gutierrez C, Robbes R (2013) *WEON: Towards a Software Ecosystem Ontology*. International Workshop on Ecosystem Architectures, pp. 16–20.

Hendricks DE (1985) Portable electronic detector system used with inverted-cone sex pheromone traps to determine periodicity and moth captures. *Environmental Entomology* 14(3): 199–204. https://doi.org/10.1093/ee/14.3.199

Hendricks DE (1990) Electronic system for detecting trapped boll weevils in the field and transferring incident information to a computer. Southwestern Entomologist 15(1): 39–48.

Holguin GA, Lehman BL, Hull LA, Jones VP, Park J (2010) Electronic traps for automated monitoring of insect populations. *IFAC Proceedings* 43(26): 49–54. https://doi.org/10.3182/20101206-3-JP-3009.00008

IAASTD (2008) *Agriculture on a Cross Road.* Executive Summary of the Synthesis Report. IAASTD Intergovernmental Plenary in Johannesburg South Africa.

Islam N, Abbasi AZ, Shaikh ZA (2010) *Semantic Web: Choosing the Right Methodologies, Tools and Standards.* International Conference on Internet and Emerging Technologies, pp. 1–5.

James A, Nair AS, Joseph D (2019) CropSense–a smart agricultural system using IoT. *Journal of Electronic Design Engineering* 5(3): 1–7.

Jiang JA, Lin TS, Yang EC, Tseng CL, Chen CP, Yen CW, Zheng XY, Liu CY, Liu RH, Chen YF, Chang WY (2013) Application of a web-based remote agro-ecological monitoring system for observing spatial distribution and dynamics of Bactrocera dorsalis in fruit orchards. *Precision Agriculture* 14(3): 323—42. https://doi.org/10.1007/s11119-012-9298-x

Kogan M, Jepson P (Eds.) (2007) *Perspectives in Ecological Theory and Integrated Pest Management.* Cambridge University Press.

Kondo T, Tsinoremas NF, Gloden SS, Johnson CH, Kutsuna S, Ishiura M (1994). Circadian clock mutants of cyanobacteria. *Science* 266: 1233–1236.

Lima MCF, de Almeida Leandro MED, Valero C, Coronel LCP, Bazzo COG (2020) Automatic detection and monitoring of insect pests—a review. *Agriculture* 10(5): 161. https://doi.org/10.3390/agriculture10050161

Liu K, Hogan WR, Crowley RS (2011) Natural language processing methods and systems for biomedical ontology learning. *Journal of Biomedical Informatics* 44: 163–179.

Liu M, Li C, Cao C, Wang L, Li X, Che J, Yang H, Zhang X, Zhao H, He G, et al. (2021) Walnut fruit processing equipment: academic insights and perspectives. *Food Engineering Reviews* 13: 822–857.

Liu P, Hong Peng Z (2013) Smart cities in China, computer. *IEEE Computer Society Digital Library* 47: 72–81.

Makokha SI, Mutoko MC, Ondieki CM (2019) Climate-smart agriculture and its contribution to sustainable livelihoods: a case study of smallholder farmers in Kenya. *Journal of Agricultural Science and Technology A* 9(9): 427–437.

Opara U (2004) Emerging technological innovation triad for smart agriculture in the 21st century part-I prospects and impacts of nanotechnology in agriculture. *Agricultural Engineering International: The CIGR Journal of Scientific Research and Development.* Invited Overview Paper V: VI.

Preti M, Verheggen F, Angeli S (2021) Insect pest monitoring with camera-equipped traps: strengths and limitations. *Journal of Pest Science* 94(2): 203–217. https://doi.org/10.1007/s10340-020-01309-4

Rassati D, Faccoli M, Chinellato F, Hardwick S, Suckling DM, Battisti A (2016) Web-based automatic traps for early detection of alien wood-boring beetles. *Entomologia Experimentalis et Applicata* 160(1): 91–95. https://doi.org/10.1111/eea.12453

Shaked B, Amore A, Ioannou C, Valdés F, Alorda B, Papanastasiou S, Goldshtein E, Shenderey C, Leza M, Pontikakos C, Perdikis D, Tsiligiridis T, Tabilio MR, Sciarretta A, Barceló C, Athanassiou C, Miranda MA, Alchanatis V, Papadopoulos N, Nestel D (2018) Electronic traps for detection and population monitoring of adult fruit flies (Diptera: Tephritidae). *Journal of Applied Entomology* 142(1–2): 43–51. https://doi.org/10.1111/jen.12422

Ünlü L, Akdemir B, Ögür E, Şahin İ (2019) Remote monitoring of European Grapevine Moth, Lobesia botrana (Lepidoptera: Tortricidae) population using camera-based pheromone traps in vineyards. *Turkish Journal of Food and Agriculture Sciences* 7(4): 652–657. https://doi.org/10.24925/turjaf.v7i4.652-657.2382

Wang N, Zhang N, Wang M (2006) Wireless sensors in agriculture and food industry recent development and future perspective. *Computers and Electronics in Agriculture* 50: 1–14.

Wark T, Corke P, Sikka P, Klingbeil L, Guo Y, et al. (2007) Transforming agriculture through pervasive wireless sensor networks. *IEEE Pervasive Computing* 6(2): 50–57.

Yang Z, Rao MN, Elliott NC, Kindler SD, Popham TW (2009) Differentiating stress induced by green bugs and Russian wheat aphids in wheat using remote sensing. *Computers and Electronics in Agriculture* 67: 64–70.

Yukikazu M, Slamet U, Keita H, Kazuhiro S (2012) Proposed of cultivation management system with informatics and communication technology. *1st IEEE Global Conference on Consumer Electronics*, pp. 175–176.

Chapter 5

Methodological approaches for smart agriculture and its applications

Zainab Fatima, Muhammad Farhan, Ghulam Mohiuddin, Muhammad Zaid, Mohd Usman Khan, and Muhammed Muaz

5.1 INTRODUCTION

Smart agriculture is an emerging concept that seeks to improve traditional agricultural practices by leveraging the use of technology. It combines the use of information and communication technologies (ICTs), such as sensors, big data, artificial intelligence, and robotics, with traditional farming methods to create a more efficient and sustainable agricultural sector [1]. Smart agriculture has the potential to improve farming productivity, profitability, and sustainability. Technology improvements are having a significant impact on many aspects of life. People have been striving toward automation technologies with some level of replacement or reduction of the human component in processes for many years. According to current trends, the next era will revolve around the ideas of ubiquitous computing and context-aware computing. Smart agriculture is a rapidly evolving field that is transforming the way traditional agriculture is practiced. By using innovative technologies it is possible to monitor, predict, and optimize crop production (Figure 5.1). Smart agriculture offers several benefits, such as improved soil health, increased yields, reduced water usage, and improved pest management [2]. It can also improve the safety and efficiency of farm operations, reduce environmental impacts, and enhance the quantity as well as the quality of agricultural products [3].

5.1.1 Overview of smart agriculture

A contemporary farming approach, "smart agriculture" makes use of cutting-edge technologies and techniques to increase agricultural efficiency and productivity while reducing the environmental impact of farming. Smart agriculture combines different technologies, such as automation and robotics, drones and sensors, machine learning, and artificial intelligence, to monitor, analyze, and manage agricultural processes effectively. It is the future of farming as it ensures sustainable production, cost-effectiveness, and food

DOI: 10.1201/9781003495314-5

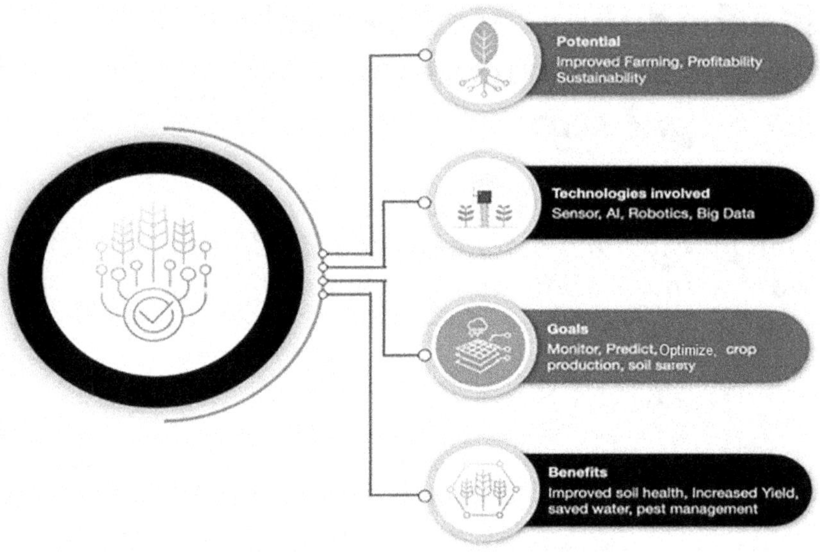

Figure 5.1 Conceptual Architecture of Smart Agriculture.

security. It helps farmers make data-driven decisions to optimize production and crop yields while minimizing resource wastage. Smart agriculture is an essential step toward achieving global food security and sustainability.

5.1.2 Challenges faced by the agricultural sector

The agricultural sector faces several challenges that threaten its sustainability and food security. By 2050, the world's population is projected to increase by an astounding 9.7 billion people. The need for food will rise as a result of population growth. Meeting this rising demand while reducing farming's impact on the environment is the problem facing the agricultural sector. Despite the potential of smart agriculture, the agricultural sector faces several challenges. These include climate change, soil degradation, water scarcity, and pest infestations. Additionally, farmers face declining profits due to rising input costs and declining prices for agricultural products [4]. Furthermore, the lack of access to technology and the lack of training in its use can limit the adoption of smart agriculture practices [5]. Finally, the lack of infrastructure and data networks can limit the effectiveness of smart agriculture [6]. These challenges require innovative solutions and technologies to ensure sustainable and productive farming practices. Smart agriculture is a combination of different technologies and techniques that are used to optimize agricultural practices (Figure 5.2).

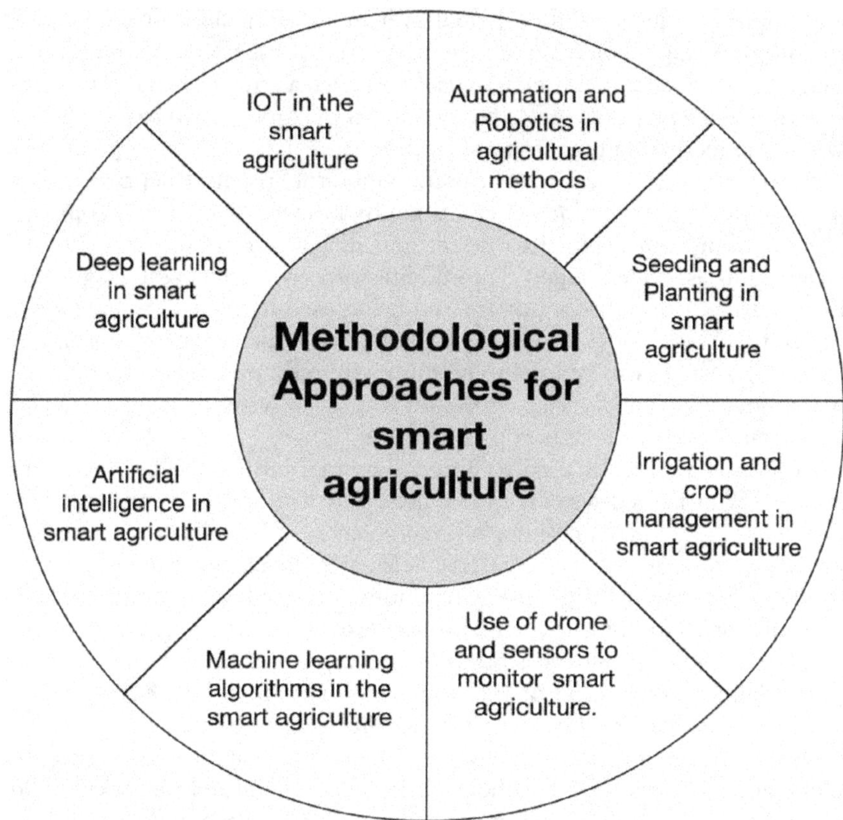

Figure 5.2 Methodological Approaches for Smart Agriculture.

5.1.3 Automation and robotics

Automation and robotics are two of the most important elements of smart agriculture. They play crucial roles in smart agriculture. Automation and robotics have transformed the way farmers manage their farms. They can perform repetitive tasks, such as planting, weeding, and harvesting, with high precision and accuracy. The use of these technologies is becoming increasingly popular in agricultural production, as they allow for improved efficiency and accuracy in the management of crops and livestock. This technology reduces the workload of farmers and improves production efficiency, making it possible to manage larger farms with fewer labour requirements.

The use of automation and robotics in agriculture also increases crop yields, reduces crop damage, and minimizes resource wastage. The adoption of automation and robotics in agriculture has the potential to revolutionize the agricultural sector by increasing productivity, reducing labour costs, and improving sustainability. Robotics in smart agriculture has been used to automate the tedious and labor-intensive tasks associated with crop and livestock production. Automated machines such as tractors, harvesters, and soil samplers have made it feasible to shorten the time needed to finish a task while also increasing precision and accuracy.

Automated systems can also be used to monitor crop health, provide diagnostics, and make decisions in real time. By utilizing sensors and computer vision, robots can be used to detect and diagnose crop diseases, identify weeds, and estimate yields [7]. In addition to crop and livestock production, automation and robotics are also being used in other aspects of smart agriculture, such as precision irrigation and water management. Automated irrigation systems can be used to monitor soil moisture levels and adjust the water flow accordingly. This allows for improved water efficiency and conservation, as well as better crop yields [8].

Robotics can also be used to improve animal husbandry and feed management. Automated systems can be used to detect the health and nutrition of livestock, as well as monitor animal behavior. This allows for improved animal welfare, as well as increased efficiency in the management of feed [9]. In the future smart, agriculture will continue to rely heavily on automation and robotics. The ability to implement automated systems on a bigger scale will become increasingly possible as technology advances and costs drop. In particular, the deployment of autonomous drones and robots has the potential to completely alter how agricultural operations are run.

Weed and pest management, soil sampling, and crop reconnaissance can all be done by autonomous robots. Drones can be utilized for tasks including crop monitoring, irrigation control, and fertilizer administration [10]. The use of automation and robotics in smart agriculture also has the potential to improve food safety and quality. Automated systems can be used to monitor food quality in real time, as well as detect and prevent food contamination. By utilizing data-driven analytics, automated systems can also be used to optimize the quality and yield of crops and livestock [11]. For an understanding of the various issues in the agricultural world handled by autonomous devices, Table 5.1 presents a detailed overview of automation and robotics in smart agriculture.

5.1.4 Seeding and planting in smart agriculture

Seeding and planting are critical processes in agriculture that determine crop yields and productivity. The selection of seeds is a critical step in the process of planting and harvesting crops. The right seed varieties should be chosen based on the soil type, climate, and desired yield. It is important to select varieties that are resilient to pests and disease and are adapted to the local environmental conditions. Furthermore, the amount of seed used should be determined according to the expected yield [12].

Table 5.1 A Detailed Overview of Automation and Robotics in Smart Agriculture

Sl. no.	*Problem*	*Platform*	*Components*	*System control*
1	Crop Monitoring	Unmanned Aerial Vehicles (UAVs	Multi-spectral cameras, LIDAR, GPS	Autonomous flight control, image processing
2	Soil Analysis	Ground-based Robots	Soil sensors, spectrometers	Automated soil sampling, real-time data analysis
3	Crop Harvesting	Autonomous Harvesting Robots	Robotic arms, sensors	Automated navigation, object detection and picking
4	Irrigation	Automated Irrigation Systems	Soil moisture sensors, weather data, water pumps	Automated control based on real-time data, precision irrigation
5	Pest Control	Autonomous Pest Detection and Eradication Robots	Cameras, sensors, spraying mechanisms	Automated pest detection and targeted spraying
6	Livestock Monitoring	Wearable Devices and Sensors	GPS, accelerometers, heart rate monitors	Real-time monitoring, automated health alerts
7	Greenhouse Management	Automated Greenhouse Systems	Climate sensors, CO2 sensors, lighting systems	Automated climate control, remote monitoring
8	Autonomous Tractors	Autonomous Tractors	GPS, obstacle detection sensors, steering systems	Autonomous navigation and operation
9	Sorting and Grading	Automated Sorting and Grading Systems	Computer vision cameras, conveyor belts, sorting algorithms	Automated sorting and grading based on quality and size
10	Weeding	Autonomous Weeding Robots	Cameras, sensors, weeding mechanisms	Automated weed detection and targeted weeding

Precision seeding and planting techniques ensure that crops are planted at the right depth and spacing, which promotes optimal growth and reduces resource wastage. Technologies, such as precision planting, can optimize seed placement, depth, and spacing to improve crop yields and minimize resource wastage. Soil moisture sensors and GPS technology can also be used to determine the optimal time for planting and to monitor soil conditions, which ensure that crops are planted in the best possible conditions. By adopting precision seeding and planting techniques, farmers can improve crop yields, reduce resource wastage, and increase sustainability. To ensure the best results, soil preparation is an important step before planting. The soil should be tested for fertility and other characteristics such as pH value, texture, and weed pressure. The soil should be amended to ensure optimal conditions for crop growth. In addition, proper irrigation, drainage, and nutrient management should be implemented to promote optimal crop growth [13].

Timely and accurate planting is essential for good crop yields. Planting should be done when the soil conditions are optimal, and the right amount of seed should be planted at the correct depth. Planting at the right time of the year can also help to reduce losses due to pests and disease [14]. Harvesting is the final step in the process of smart agriculture and should be carried out when the crop is at its peak. The right harvesting techniques should be employed to ensure the highest yield and quality of the crop. Proper post-harvest handling is also important to ensure that the crop remains in good condition until it is sold [15].

5.1.5 Irrigation and crop management in smart agriculture

Irrigation and crop management are essential for achieving sustainable and productive agriculture. Efficient irrigation systems and crop management techniques can reduce water consumption, minimize nutrient loss, and increase crop yields. The use of sensors and remote monitoring systems can help farmers determine the optimal irrigation schedule and monitor crop growth and health. Technologies, such as drip irrigation and fustigation, can improve water and nutrient management, that is, reducing water wastage and nutrient loss while increasing crop yields. The adoption of smart irrigation and crop management techniques can help farmers optimize resource usage, reduce production costs, and increase sustainability.

Water management is the process of efficiently managing the water resources available for irrigating crops. This involves selecting the most suitable irrigation methods, applying the correct amount of water, and scheduling irrigation to minimize water waste and optimize crop yield [16]. Different methods of irrigation include surface irrigation, subsurface irrigation, and sprinkler irrigation [17].

Surface irrigation is the most common and economical method of irrigation but is not suitable for all crops. Subsurface irrigation is suitable for crops that require a high water table, and sprinkler irrigation is suitable for irrigated crops that require frequent and precise amounts of water. Crop selection is important for successful smart agriculture. The selection of suitable crops for the local environment and soil type is essential for optimizing crop yields [18].

Soil type, climate, and land-use patterns should be taken into account when selecting crops. Additionally, the local market and consumer preferences should be considered when selecting crops. The use of precision agriculture techniques, such as remote sensing and geographic information systems (GIS), can help farmers to identify the most suitable crops for their local environment [19].

The use of modern technology can help farmers optimize water management and crop selection. Automated systems and sensors can be used to monitor soil moisture and water levels and to automate irrigation systems [20]. Decision support systems can help farmers to make informed decisions regarding irrigation and crop selection [21]. Additionally, satellite imagery and GIS can be used to monitor crop growth and detect potential problems in the field [22].

5.1.6 Drones and sensors

Smart agriculture is a technology-driven approach to agricultural production that combines the use of drones and sensors to optimize crop production [23]. Drones and sensors enable precision agriculture and provide farmers with real-time insights into their operations. Drones, which are unmanned aerial vehicles (UAVs), have the potential to revolutionize precision agriculture, as they provide farmers with an effective tool to monitor, assess, and manage their crops [24].

Drones and sensors have revolutionized the agricultural sector by providing farmers with real-time data on the crop to optimize irrigation scheduling; sensors can provide precise information on soil moisture content, temperature, humidity, and other environmental factors. For instance, using soil moisture sensors, farmers can ensure that their crops receive the optimal amount of water for their growth and development. Drones can also play a significant role in monitoring crop growth and health, as they can capture high-resolution images of crops and provide information on plant health and stress.

The data collected by sensors and drones can be analyzed using machine learning algorithms and AI-based systems to provide farmers with actionable insights on crop growth, yield, and potential problems. Sensors are also essential for successful precision agriculture. They enable farmers to measure and monitor soil and crop health, as well as the presence of weeds, pests, and nutrient deficiencies [25]. They can also detect water, pH, and temperature levels in fields and provide real-time data about weather conditions, such as wind speed, air temperature, and humidity [26].

This data can be used to make more informed decisions regarding irrigation, fertilization, and pest control. The combination of drones and sensors in smart agriculture can help farmers improve crop yield, reduce inputs, and increase efficiency [27]. Farmers may better understand their farms and manage their resources by employing drones to collect more data. Additionally, sensors can be used to monitor crop health and detect potential problems quickly, allowing farmers to respond quickly and proactively [28].

5.1.7 Machine learning algorithms in smart agriculture

Machine learning algorithms are being increasingly employed in agriculture to automate decision-making processes and optimize crop yield. Machine learning can be used to predict crop yield based on factors such as weather patterns, soil quality, and the use of fertilizers and pesticides. This technology can also be used to detect and diagnose diseases in crops, enabling farmers to take prompt measures to control their spread. Additionally, farmers can utilize machine learning algorithms to monitor the productivity and health of their animals, allowing them to improve their feeding and breeding practices for the highest possible production and quality.

Machine learning algorithms offer the potential to revolutionize the way agricultural processes are managed, providing data-driven decision-making capabilities and enabling the optimization of agricultural resources. This chapter focuses on the use of machine learning algorithms in smart agriculture, identifying the potential of these algorithms to improve the efficiency, productivity, and sustainability of agricultural processes [29].

Machine learning algorithms are used to analyze vast amounts of data collected from various sources, such as satellite imagery, soil sensors, and weather measurements, to generate insights into crop yield, soil conditions, and pest behavior [30]. This data can then be used to inform decision-making regarding the management of agricultural resources, such as irrigation, fertilization, and pest control [31]. The use of machine learning algorithms in smart agriculture has been further enabled by the emergence of cloud computing and big data analytics [32].

Cloud computing has allowed data to be analyzed in real time, enabling the rapid deployment of machine learning models to analyze large datasets and generate insights [33]. Big data analytics provides the ability to interpret and extract meaningful insights from large sets of data, enabling the development of sophisticated models that can be used to predict crop yields, detect diseases, and optimize agricultural processes [34].

5.1.8 Artificial intelligence in smart agriculture

Artificial intelligence (AI) is becoming increasingly relevant in agriculture, enabling farmers to collect and analyze data in real time, make informed decisions, and optimize crop yield. AI-powered systems can be used to optimize irrigation, fertilizer, and pesticide use, thus reducing waste and environmental impact. AI can also be used to analyze weather patterns and climate data, providing farmers with accurate predictions of weather events and enabling them to take proactive measures to mitigate potential damage.

AI-enabled technologies allow for the automation of some agricultural processes such as planting, harvesting, and irrigation, making them more efficient and reducing the need for manual labour. AI-based systems can also be used to monitor the environment, predict changes in weather patterns and crop yields, and optimize agricultural supply chains, ensuring that the appropriate resources are accessible in the appropriate settings at the appropriate time [35].

AI-enabled technologies are also being used to improve the traceability of food and create transparent supply chain networks. This allows for better monitoring of food production and better control of food safety. Additionally, AI-based systems can be used to monitor the environment and provide recommendations to reduce the environmental impact of food production [36]. Predictive models can be developed using AI-based systems to help farmers understand better how climate change would affect their animals and crops. AI-based systems can be used to make predictions of crop yields and weather patterns, allowing for better preparedness and mitigation of climate-related risks [37].

However, the use of AI in agriculture also poses ethical concerns, particularly concerning privacy, data security, and bias in decision-making. Farmers must be mindful of the potential negative impacts of AI, such as the loss of jobs and increased dependency on technology. Therefore, it is essential to implement appropriate ethical frameworks and regulations to ensure the responsible use of AI in agriculture. An application of AI in farming can be overviewed in Figure 5.3

5.1.9 Deep learning in smart agriculture

Artificial neural networks are used in the field of deep learning (DL), a branch of machine learning, to analyze huge and complicated information. DL has been used in agriculture for a variety of tasks, including crop categorization, yield forecasting, and disease detection. For instance, DL algorithms can be used to classify crops based on their visual appearance, enabling farmers to identify and monitor different plant species and their growth patterns. DL can also be used to predict crop yield based on various factors such as weather, soil quality, and the use of fertilizers and pesticides.

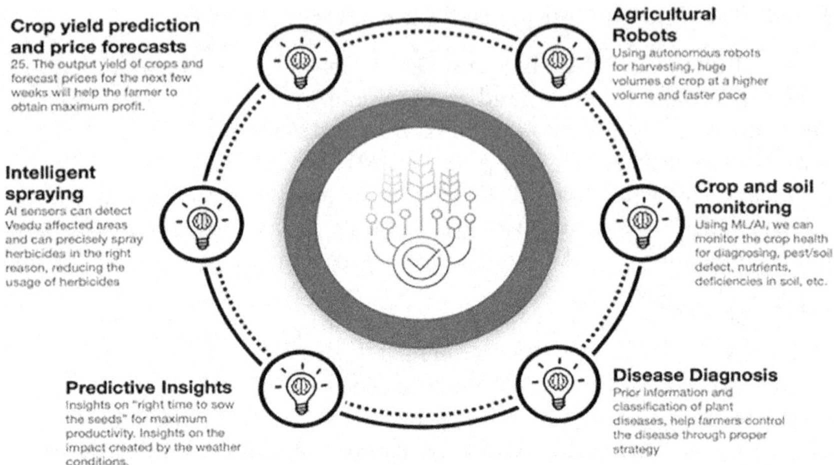

Figure 5.3 Application of AI in Smart Agriculture.

DL is one of the most promising techniques in this field. It has the potential to revolutionize the way crops are grown, harvested, and managed [38]. DL techniques can be utilized for various applications such as crop monitoring, precision farming, yield prediction, crop disease diagnosis, and pest control [39]. Crop data collection, processing, and analysis could be revolutionized by DL. It can be useful to pinpoint areas of crop production that need to be improved and to create plans for raising crop yield and maximizing resource use [40].

DL can also be used to monitor crop growth, detect crop diseases and pests, and predict crop yield. Furthermore, they can be used to generate accurate and timely crop yield forecasts, in turn informing decisions about planting and harvesting and developing decision support systems for precision agriculture, which can be used to optimize crop management and reduce costs associated with crop production [41].

One of the major challenges in applying DL to agriculture is the need for large and diverse datasets that accurately capture the variability and complexity of agricultural systems. Additionally, DL algorithms require significant computational resources and may be limited by hardware and software constraints. Therefore, it is essential to optimize the design and implementation of DL algorithms for efficient and effective use in agriculture. For an understanding of the various issues in the agricultural world handled by autonomous devices, Table 5.2 presents a comprehensive machine learning algorithms and deep learning in smart agriculture.

5.1.10 IoT in smart agriculture

The agricultural industry has been transformed by the Internet of Things (IoT). It enables farmers to make data-driven decisions and optimize their

Table 5.2 A Comprehensive Table of Machine Learning Algorithms and Deep Learning in Smart Agriculture

Sl. no.	*Algorithm/technique*	*Data*	*Accuracy*	*Applications in agriculture*
1	Linear Regression	Yield data from a corn field	R-squared value of 0.82	Yield prediction, optimal harvest time determination
2	Logistic Regression	Images of tomato leaves with and without bacterial spot disease	95% accuracy	Disease detection, early intervention for crop protection
3	Decision Trees	Soil quality data from a vineyard	87% accuracy	Soil quality analysis, vineyard management
4	Random Forest	Images of apple trees with and without apple scab disease	92% accuracy	Disease detection, early intervention for crop protection
5	Support Vector Machines (SVM)	Hyperspectral data from a wheat field	93% accuracy	Crop classification, yield prediction, soil quality analysis
6	Neural Networks	Data on plant height, leaf area, and chlorophyll content from a tomato greenhouse	R-squared value of 0.95	Plant growth prediction, yield prediction
7	Convolutional Neural Networks (CNN)	Images of grapevines with and without powdery mildew disease	96% accuracy	Disease detection, early intervention for crop protection
8	Recurrent Neural Networks (RNN)	Weather data from a maize field	Mean absolute error of 0.31	Weather forecasting, crop yield prediction
9	Long Short-Term Memory (LSTM)	Soil moisture data from a potato field	Mean squared error of 0.003	Irrigation management, optimal harvest time determination
10	Generative Adversarial Networks (GAN)	Images of wheat fields from different seasons	Realism score of 8.7 out of 10	Synthetic data generation, yield prediction

crop yields. Smart agriculture is an application of IoT in which farmers use sensors, devices, and data analytics to manage crops and livestock efficiently. Real-time crop and soil condition monitoring has been made possible by the implementation of IoT in agriculture, thus providing accurate and timely information for decision-making. It is an amalgamation of various information, communication, and sensing technologies that integrate data from various sources, such as real-time weather, soil, water, and air quality, to provide farmers with timely and accurate information to support decision-making [42].

One of the significant advantages of IoT in smart agriculture is precision farming. Precision farming uses various technologies such as sensors, GPS, drones, and machine learning algorithms to monitor crops' growth, health, and productivity. This approach allows farmers to adjust the amount of water, fertilizer, and pesticides needed for each crop, thus reducing wastage and increasing yields. Additionally, IoT sensors can also help in monitoring soil conditions, such as temperature, moisture, and pH levels, enabling farmers to optimize the use of fertilizers and other inputs. The application of IoT in smart agriculture is livestock management. The location, behavior, and health of animals can be tracked using IoT devices like RFID systems, sensors, and cameras [44].

For instance, sensors attached to cows can track their activity levels, feeding patterns, and milk production, enabling farmers to make informed decisions about their health and nutrition. Similarly, RFID tags can be used to track the location and movement of livestock, reducing the risk of theft and improving overall management efficiency. IoT can also be used in precision irrigation, which is essential for areas where water resources are scarce. Smart irrigation systems use IoT sensors and weather data to determine the optimal amount of water required for each crop. These systems can also be used to monitor soil moisture levels, thus preventing overwatering and reducing water wastage. By using IoT in precision irrigation, farmers can reduce water usage by up to 30%, resulting in significant cost savings [43].

Furthermore, IoT can also help in predicting weather conditions and natural disasters, which is crucial for farmers. IoT sensors can be used to monitor weather patterns, such as temperature, humidity, and wind speed, allowing farmers to decide when and how to grow and harvest and fertilize their crops with knowledge. Furthermore, IoT can be used to forecast natural calamities like floods, storms, and wildfires, enabling farmers to take timely action and minimize the damage to their crops. IoT has tremendous potential in smart agriculture, enabling farmers to optimize their crop yields, reduce wastage, and improve overall efficiency. There is still much room for innovation and progress in IoT integration in agriculture, although it remains in its early phases.

5.2 SMART FRAMEWORK BASED ON FARMERS' PSYCHOLOGY

The agricultural industry has evolved considerably over the last few decades, yet the psychological well-being of farmers remains a largely overlooked aspect. This book chapter aims to explore the potential of a SMART framework based on farmers' psychology to improve their overall well-being and productivity. The framework consists of five components: Self-awareness, Mindfulness, Attitude, Reflection, and Training. It can be overviewed in Figure 5.4.

- Self-awareness refers to the ability to recognize one's own emotional and cognitive states and is the foundation of the framework.
- Mindfulness is the practice of being present and non-judgmental at the moment.
- Attitude is how one perceives and responds to a situation.
- Reflection is the process of thinking back on one's feelings and thoughts to obtain understanding and insight.
- The process of learning new information and abilities to enhance performance is known as training [45].

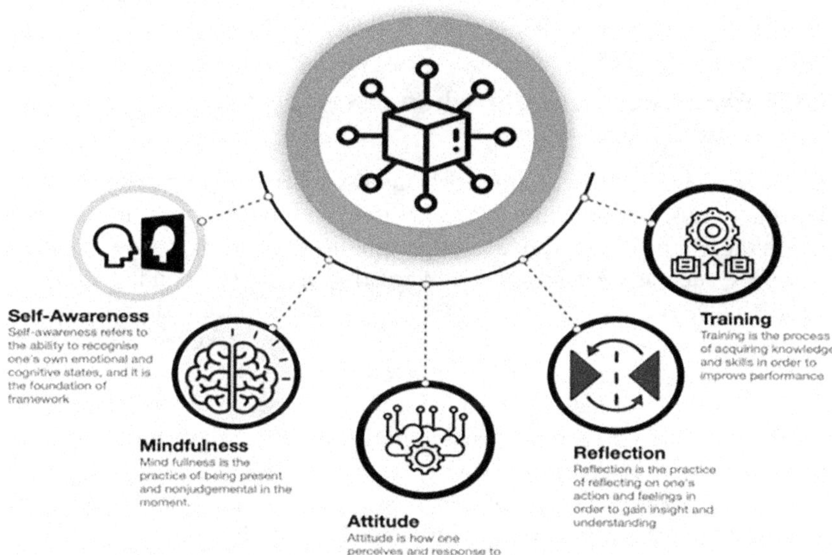

Figure 5.4 Components of Smart Framework.

5.2.1 Overview of smart framework

Smart agriculture is revolutionizing the way we approach agriculture, and one of the key drivers for smart agriculture is the development of smart frameworks based on farmers' psychology. A smart framework is a decision support system that helps farmers make informed decisions based on various factors such as crop data, weather data, and farmers' psychology [46]. By integrating farmers' psychology into the smart framework, the system can provide personalized recommendations and solutions that take into account the individual farmer's preferences, beliefs, and attitudes toward farming.

The technological acceptance model (TAM) is one of the most significant models of planning in the environment of smart agriculture. According to TAM, for a user to embrace and adopt a technology, it must first appear beneficial and simple to use. Furthermore, TAM suggests that social influences, such as perceived peer pressure and perceived social support, can also influence a user's decision to accept and adopt new technologies [47].

The smart framework based on farmers' psychology is designed to help farmers in several areas, including crop management, irrigation, and pest control. This framework can help farmers identify the most effective techniques to use, as well as the best times to use them. With this framework, farmers can also monitor their crops more effectively, identifying and addressing issues before they become major problems. Additionally, the framework can help farmers increase their productivity and profitability by providing recommendations for optimal crop rotation and planting schedules, as well as identifying the most profitable crops for their region and market.

5.2.2 Factors to consider when developing smart framework

When developing a smart framework based on farmers' psychology, numerous things must be taken into account. First and foremost, it's critical to recognize the key psychological factors that influence farmers' decision-making processes, such as their attitudes toward risk, innovation, and technology. Additionally, cultural and social factors should also be considered, as they can significantly impact farmers' willingness to adopt new technologies. An important factor to consider when developing a smart framework is the type of data that will be used to inform the system. Many kinds of details, such as weather information, can be included in this data, soil data, and crop data, which can be collected using a variety of sensors [48].

Machine learning algorithms can then be used to analyze this data and provide personalized recommendations for farmers. It is essential to design the smart framework in such a way that it is accessible and user-friendly for farmers. Many farmers may not have extensive technical knowledge or experience with technology. Thus, the framework needs to be created in a form that is simple to use and comprehend. Making sure the framework is affordable and sustainable is crucial so that it can be adopted by farmers across different socioeconomic backgrounds.

5.2.3 Benefits of smart framework

There are numerous benefits from developing a smart framework based on farmers' psychology. First, it can help farmers make more informed decisions, leading to improved crop yields, higher profits, and reduced costs. By providing personalized recommendations based on individual farmers' preferences, beliefs, and attitudes, the framework can help farmers achieve their goals more efficiently. A smart framework can help farmers adopt sustainable and environment-friendly practices, such as precision agriculture and precision irrigation. This can help reduce the negative impact of agriculture on the environment and promote the development of more sustainable farming practices [49].

5.3 SUMMARY

Smart agriculture is a rapidly growing field, motivated by the necessity for efficient, affordable, and sustainable agriculture methods to meet the rising needs of an increasing world population. In this chapter, we explored various methodological approaches for smart agriculture, including automation and robotics, seeding and planting, irrigation and crop management, drones and sensors, machine learning algorithms, artificial intelligence, and deep learning. These approaches can help farmers optimize their farming practices by reducing manual labour, minimizing resource wastage, and increasing yields. Moreover, we discussed the smart framework based on farmers' psychology, which can be developed by considering various factors, including farmers' behavior, climate, soil, and crop data. By integrating all these factors, farmers can optimize their practices and achieve better yields. The benefits of smart agriculture are significant, including reduced costs, higher yields, and sustainable practices that benefit the environment [50].

5.4 IMPLICATIONS FOR FUTURE RESEARCH

The potential of smart agriculture includes enormous issues that must be resolved. The necessity to create scalable and affordable solutions that smallholder farmers can use is one of the major problems, which are the backbone of agriculture in many countries. Additionally, there is a need for further research to optimize the algorithms and models used in smart agriculture, which can lead to more accurate predictions and better results. Further research can also focus on developing a better understanding of farmers' behavior, preferences, and decision-making processes to create more effective smart farming solutions.

5.5 FUTURE DIRECTIONS

Looking to the future, there is an urgent need for the widespread adoption of smart agriculture practices to meet the increasing demand for food, fuel, and fibre while preserving the environment. With advancements in technology and data analytics, there is enormous potential for the development of new, innovative solutions that can help farmers increase yields, reduce resource wastage, and improve sustainability. Governments, research institutions, and private sector organizations must work together to develop and promote smart agriculture solutions that are accessible, scalable, and sustainable for all farmers. Smart agriculture can play a significant role in meeting the growing global demand for food while ensuring that the environment is protected for future generations. [51]

5.6 CONCLUSION

In conclusion, smart agriculture is a promising solution to address the challenges faced by the agricultural sector. By adopting a smart framework based on farmers' psychology and using automation, sensors, and machine learning algorithms, farmers can optimize their practices and achieve better yields while reducing costs and minimizing resource wastage. Governments, research institutions, and private sector organizations must work together to promote the widespread adoption of smart agriculture practices, and further research must focus on developing scalable, cost-effective solutions that can be adopted by smallholder farmers. With the right approach, smart agriculture can help meet the growing global demand for food while ensuring a sustainable future for all.

REFERENCES

[1] Food and Agriculture Organization of the United Nations, *Smart Agriculture*, 2020. http://www.fao.org/smart-agriculture/en/.

[2] C. Roth and J. Lammel, "Smart agriculture: digital farms and digital farmers," in G. Schlee, H. de Meer, and M. van Eupen (Eds.), *Digital Agriculture: An Interdisciplinary Approach* (pp. 3–14). Berlin, Heidelberg: Springer, 2016.

[3] D. Della, G. Vilela and G. Armano, "Smart agriculture: technologies and challenges," in *Smart Agriculture: Technologies and Challenges* (pp. 1–15). Cham: Springer, 2017.

[4] United Nations Conference on Trade and Development, *Agricultural Commodity Prices: Causes and Impacts*, 2019. https://unctad.org/en/PublicationsLibrary/webdiaepcb2019d3_en.pdf.

[5] H. Smith. *Smart Farming: The Benefits, Barriers, and Opportunities*, 2018. https://www.kauffman.org/what-we-do/resources/entrepreneurship-policy-digest/smart-farming-the-benefits-barriers-and-opportunities.
[6] Y. Fan, X. Chen, and M. Ouyang, "Big data in agriculture: a review," *Sensors*, vol. 18, no 7, p. 2260, 2018. https://doi.org/10.3390/s18072260
[7] H. N. Nguyen, B. H. Lee, and S. Y. Kim, "Robotics and automation in precision agriculture," *Robotics Autonomus Systems*, vol. 97, pp. 99–111, 2018.
[8] P. A. Calvete and S. M. Guevara, "A review of the current state of the art in precision irrigation systems," *Computers and Electronics in Agriculture*, vol. 154, pp. 5–14, 2018.
[9] M. A. Abeysekera et al., "Robotics and automation in animal husbandry: a review," *Computers and Electronics in Agriculture*, vol. 154, pp. 15–26, 2018.
[10] H. M. Ouhbi et al., "A review of automation and robotics in agriculture: current status, challenges and future perspectives," *Computers and Electronics in Agriculture*, vol. 154, pp. 27–40, 2018.
[11] S. L. Merritt et al., "Robotics and automation for food safety and quality assurance," *Trends in Food Science & Technology*, vol. 97, pp. 85–93, 2020.
[12] S. K. Pandey et al., "Seed selection for sustainable agriculture," in *Advances in Agronomy* (vol. 140, pp. 267–294). Elsevier, 2018. https://www.sciencedirect.com/science/article/pii/S0065211317300859
[13] S. K. Singh and A. Gupta, "Soil preparation for sustainable agriculture," in *Advances in Agronomy* (vol. 139, pp. 337–362). Elsevier, 2018. https://www.sciencedirect.com/science/article/pii/S0065211317300741
[14] P. H. Brown and A. J. Storer, "Timing of planting and harvesting for sustainable agriculture," in *Advances in Agronomy* (vol. 139, pp. 363–389), Elsevier, 2018. https://www.sciencedirect.com/science/article/pii/S0065211317300753, 2018
[15] M. S. Smidt and M. J. Boyett, "Harvesting for sustainable agriculture," in *Advances in Agronomy* (vol. 139, pp. 391–420). Elsevier, 2018. https://www.sciencedirect.com/science/article/pii/S0065211317300765
[16] S. Wani, C. J. Reddy, and C. M. Kijne, "Sustainable agricultural water management in India," *Current Opinion in Environmental Sustainability*, vol. 1, no. 2, pp. 135–145, 2009.
[17] S. K. Sharma and D. D. Sharma, "Irrigation methods: a brief review," *Annals of Agricultural Research*, vol. 24, no. 2, pp. 234–236, 2003.
[18] A. C. R. Rao, "Crop selection and site selection," in S. P. Mathur, (Ed.), *Handbook of Agricultural Geography & Geology* (pp. 83–91). New Delhi, India: Oxford & IBH, 1982.
[19] P. Rahmati, R. Sarrafzadeh, and M. Farsi, "A review of precision agriculture techniques and applications," *Computers and Electronics in Agriculture*, vol. 83, pp. 37–45, 2012.
[20] S. G. González-Villa, A. P. Córdoba-García, and E. M. Sánchez-González, "Automation systems for irrigation management in precision agriculture," *Computers and Electronics in Agriculture*, vol. 83, pp. 46–54, 2012.
[21] S. S. D. V. S. Rao and M. S. Chari, "Decision support systems in agricultural management," *Decision Support Systems*, vol. 34, no. 2, pp. 181–190, 2003.
[22] G. M. Foody, "Status of land cover classification accuracy assessment," *Remote Sensing of Environment*, vol. 80, no. 1, pp. 185–201, 2002.
[23] J. P. Hurley, "Precision agriculture: the application of drones and sensors," *Computers and Electronics in Agriculture*, vol. 138, pp. 90–98, 2017.

[24] J. Stroud, et al., "Drone-based crop scouting for precision agriculture," *Computers and Electronics in Agriculture*, vol. 125, pp. 90–98, 2016.
[25] M. A. Anjum, et al., "Sensors in precision agriculture: a review," *Computers and Electronics in Agriculture*, vol. 158, pp. 112–122, 2019.
[26] K. Gebrehiwot, et al., "Application of precision agriculture technology for crop production: a review," *Computers and Electronics in Agriculture*, vol. 126, pp. 246–255, 2016.
[27] R. Rai, et al., "Precision agriculture: a review on applications of remote sensing and GIS," *International Journal of Applied Engineering Research*, vol. 12, no. 4, pp. 1285–1290, 2017.
[28] Sánchez, J., et al., "Smart agriculture: a review of the application of IoT and wireless sensor networks," *Sensors*, vol. 17, no. 3, p. 523, 2017.
[29] M. K. Kaul, A. K. Verma, and R. Sharma, "Machine learning algorithms: a survey," in *2020 6th International Conference on Advanced Computing & Communication Systems (ICACCS)* (pp. 156–161). IEEE, 2020.
[30] P. S. Khedkar, A. S. Bhosale, S. K. Shinde, and S. M. Kale, "A review of smart agriculture: a new era of modern farming," *International Journal of Recent Technology and Engineering*, vol. 8, no. 1, pp. 1741–1745, 2019.
[31] G. P. Kouroupetroglou, S. M. Poulos, D. Koutsoyiannis, and J. K. Papanicolaou, "Modeling agricultural decision-making using machine learning algorithms," *Hydrological Sciences Journal*, vol. 63, no. 10, pp. 1551–1562, 2018.
[32] M. Farhan, M. Anas, and M. Azeem. "Rolling barriers: emerging concept to reduce road accidents: An Indian perspective," in *IOP Conference Series: Materials Science and Engineering* (vol. 404, no. 1, p. 012045). IOP Publishing. 2018. https://doi.org/10.1088/1757-899X/404/1/012045.
[33] O. Olaleye and S. S. Onifade, "Big data analytics and cloud computing in agriculture: a review," *Cogent Engineering*, vol. 5, no. 1, p. 1535442, 2018.
[34] S. A. Salazar, R. T. N. Costa, and M. G. V. Silva, "Cloud computing and big data analytics for agriculture and forestry applications: a review," *Computers and Electronics in Agriculture*, vol. 158, pp. 103542–103542, 2020.
[35] C. M. Reif and A. M. Grigoriu, "Precision agriculture: a review of machine learning methods," *Computers and Electronics in Agriculture*, vol. 162, p. 103976, 2020.
[36] J. Li et al., "Artificial intelligence in agriculture: challenges and opportunities," *International Journal of Agricultural and Biological Engineering*, vol. 11, no. 4, pp. 1–7, 2018.
[37] J. D. Díaz et al., "Smart agriculture: trends and opportunities for the use of artificial intelligence," *Sensors*, vol. 18, no. 5, pp. 1445–1456, 2018.
[38] S. Sharma et al., "Use of artificial intelligence models in agriculture: A review," *Computers and Electronics in Agriculture*, vol. 162, pp. 18–25, 2019.
[39] S. Karp and A. Zaslavsky, "Deep learning for agriculture: current applications and research opportunities," *Computers and Electronics in Agriculture*, vol. 149, pp. 11–25, 2018.
[40] S. Nagesh and U. Desai, "Smart agriculture using IoT and deep learning," *International Journal of Computer Applications*, vol. 167, no. 19, pp. 16–20, 2018.
[41] N. K. Singh, S. Chaudhary, and P. K. Dwivedi, "Internet of Things (IoT) in agriculture: a comprehensive review," *International Journal of Advanced Research in Computer Science and Software Engineering*, vol. 8, no. 5, pp. 1072–1085, May 2018.

[42] K. A. Adami, F. B. R. Ramos, F. P. Pereira, and M. A. L. Oliveira, "Precision agriculture and water management: a review," *Agricultural Water Management*, vol. 189, pp. 1–11, July 2017.
[43] M. Farhan, M. T. Mastura, S. P. Ansari, M. Muaz, M. Azeem, and S. M. Sapuan, "Advanced potential hybrid biocomposites in aerospace applications: a comprehensive review," in *Advanced Composites in Aerospace Engineering Applications*. Cham: Springer Nature, 2022. https://doi.org/10.1007/978-3-030-88192-4_6.
[44] G. Wang, S. Dong, and J. Xu, "A comprehensive review on the application of machine learning and artificial intelligence in agriculture," *Computers and Electronics in Agriculture*, vol. 173, pp. 104955, May 2020.
[45] P. Williams and S. Robinson, "The impact of the SMART framework," in M. Smith (Ed.), *Handbook of Farming Psychology*, Elsevier, 2020.
[46] A. Moghaddam, S. A. M. Sezavar, and M. Moazami, "The role of perceived support in farmers' adoption of precision agriculture technology in Iran," *Computers and Electronics in Agriculture*, vol. 154, pp. 154–164, 2018.
[47] J. E. O. Rego, "Smart agriculture: a framework for technology adoption and implementation by agricultural producers," in E. K. F. Chan and J. Lee (Eds.), *Smart Agriculture: Technologies and Tools for Sustainable Farming* (pp. 21–40). IGI Global, 2019.
[48] K. Gupta, G. Maurya, and R. Srivastava, "A review of smart agriculture: trends, opportunities, and challenges," *Computers and Electronics in Agriculture*, vol. 181, p. 105598, 2020.
[49] J. L. García et al., "Smart agriculture: a review of the current state of the art and future research directions," *Computers and Electronics in Agriculture*, vol. 184, pp. 102971, 2020.
[50] J. Wang et al., "Smart agriculture: a review of the current state of knowledge and future research directions," *Computers and Electronics in Agriculture*, vol. 153, pp. 277–286, 2018.
[51] J. L. García et al., "Smart agriculture and precision farming: a review of the current state of the art and future research directions," *Sustainability*, vol. 11, pp. 1–20, 2019.

Chapter 6

Smart agriculture in the context of the fourth industrial revolution

Opportunities and challenges for Bangladesh

Md. Shawan Uddin, Md. Sohel Rana, and Md. Mahedi Hasan

6.1 INTRODUCTION

The Fourth Industrial Revolution (4IR), commonly known as Industry 4.0, has begun for the civilization. This twenty-first century's revolution is attributable to the convergence of automation, machine learning, robotics, artificial intelligence, and the Internet of Things (IoT). These technological advancements have recently started transitioning to the farming sector as well. 4IR is a fusion of physical, cyber, and biological systems that involves machine learning, artificial intelligence, and access to information. 4IR, which is based on cyber-physical system (CPS) technology, has improved the quality of life and the economy for all people (Xu *et al.*, 2018). CPS has the ability to sense, compute, act, and communicate, thus giving data, information, and services with quick access to pertinent information, such as machine parameters, production procedures, business transactions, and product status. The integration of cutting-edge technological efforts in the agricultural sector has significantly contributed to increasing the production volume of agricultural goods while minimizing costs and effort. More specifically, an approach known as "smart agriculture" combines IoT in order to identify the environmental or manmade factors that are especially impeding agricultural cultivation and distribute them to customers (Kapoor *et al.*, 2016).

IoT contributes to the development of a digital farming system that aids farmers in practically all aspects of farming, including cultivation, growth, harvesting, and marketing. It is anticipated that from 2017 to 2022, the global smart farming market would expand at a rate of 19.3% annually, reaching $23.14 billion (GlobeNewswire, 2018). It is important to note that among all agricultural robots used in smart farming, UAV/drones are currently generating and will likely continue to produce the most revenue. A few of the key factors contributing to this significant market growth are the constant desire for improved crop yields, the increased use of information and communication technology (ICT) in farming, and the rapid climate change occurring throughout the world (FAO, 2019). The development of a

 DOI: 10.1201/9781003495314-6

smart agriculture system using contemporary technologies offers automated systems that unquestionably assist farmers in making decisions in cultivation regarding soil fertility, humidity, temperature, soil nutrition levels, use of the best water and fertilizer, actions on climate change, or crop diseases, and also help them anticipate unfavorable situations in advance. Moreover, in order to boost agricultural output and reduce overall waste, smart agriculture uses artificial intelligence (AI) in yield monitoring, forecasting, and harvesting tasks (Ayaz *et al.*, 2019).

The population of the globe is predicted to reach 9.8 billion in 2050, up around 25% from the current number (Dobele and Zvirbule, 2019). According to a 2020 estimate, the number of hungry people has increased by about 60 million over the last five years to reach close to 690 million, or 8.9% of the world's population. As the globe must produce almost 70% more food by 2050 in order to feed an expected 9 billion people, the problem of food security will only get worse (World Bank, 2020, 2022). Almost all of the aforementioned population growth is anticipated to take place in developing nations (Kopnina and Washington, 2020). According to forecasts from the Bangladesh Bureau of Statistics (BBS), the country's population would total between 210 million and 250 million people by the year 2061, with 223 million being the average. The average forecasts indicate that, if nothing else changes, the demand for food may increase by around 50% (Asaduzzaman *et al.*, 2021).

Compared to many of its Asian neighbors, Bangladesh has made significant strides in the area of food security in recent years, with more than 58.5 million people—or 36% of the total population—afflicted by mild chronic food insecurity (IPC Level 2) and 69.8 million—or 43%—classified as having no/low chronic food insecurity (IPC Report, 2022). In Bangladesh, there are nearly 35 million people who experience moderate and severe chronic food insecurity (IPC Levels 3 and 4), making up 21% of the country's total population. Of these, 11.7 million people, or 7% of the total population, experience severe chronic food insecurity (IPC Level 4), and 23.1 million people, or 14% of the total population, experience moderate chronic food insecurity (IPC CFI Level 3) (IPC Report, 2022). The report further identifies that households most at risk of IPC CFI Levels 3 and 4 are those that primarily rely on low-value and unsustainable sources of income (which frequently produce insufficient and unpredictable income), such as unskilled daily labor, marginal farming, or traditional/subsistence fishing, and reside in regions with a high frequency of shocks, such as cyclones, flash and monsoon floods, riverbank erosion, and dry spells.

In Bangladesh, apart from rice production, over the past 20 years or so, fish production has increased at a rate of about 5% annually. Livestock and poultry products are now more readily available, as well as more varied. Commercial poultry farms that produce both meat and eggs have grown significantly lately. According to one estimate, 20 million eggs are produced per day, which is a difference of more than 50% from the 7.34 billion eggs produced annually. In addition, milk output has abruptly increased by 45–46%, or roughly 30%, without any apparent change in the number of cattle. Nevertheless, the government of Bangladesh should focus more on enhancing the agricultural production so as to meet the challenges of food insecurity in global economic meltdown or at any challenging time in the coming days. Hence, this research chapter endeavors to investigate the opportunities and challenges of IoT-based smart agriculture in the context of 4IR. Further sections of this chapter will discuss a literature review, the methodology, data analysis, and the emergence of themes. The chapter ends with a concluding section.

6.2 LITERATURE REVIEW

6.2.1 Demands for smart agriculture

Significant advances have been made throughout human history to increase agricultural productivity with fewer resources and labor demands. However, despite all of these attempts, the high population rate prevented supply and demand from matching. The estimated population of the globe in 2050 is projected to reach 9.8 billion, an increase of around 25% over the estimated population in 2019 (United Nations, 2020a). Almost all of the aforementioned population growth is anticipated to take place in developing nations (United Nations, 2020b). On the other hand, it is expected that the trend of urbanization will continue at an accelerated rate, with over 70% of the world's population (now 49%) predicted to live in cities by 2050 (United Nations, 2020c). Additionally, income levels will be many times higher than they are now, which will increase the need for food, particularly in emerging nations. People in these countries will be more conscious of their diets and the quality of their food; as a result, consumer tastes may shift from wheat and grains to legumes and then, eventually, to meat. By 2050, food production should double in order to feed this larger, more urban, and wealthier population (The Guardian, 2015). In particular, the yearly production of meat must increase by more than 200 million tons to meet the demand of 470 million tons, and the present figure of 2.1 billion tons of cereal should reach nearly 3 billion tons (Baker *et al.*, 2017).

Crop production is becoming essential not only for food but also for industry; in fact, many countries' economies depend heavily on crops like cotton, rubber, and gum. Additionally, the market for bioenergy based on food crops has lately begun to expand. Prior to a decade, only the ethanol industry used 110 million tons of coarse grains, or around 10% of total global production (Baker *et al.*, 2017). Food security is in jeopardy as food crops are increasingly used for the production of biofuels, bioenergy, and other industrial purposes. The burden on already limited agricultural resources is rising as a result of these demands. Unfortunately, only a small fraction of the earth's surface is ideal for agricultural purposes due to a variety of restrictions, including temperature, climate, topography, and soil quality, and even the majority of the favorable areas are not uniform. Many new variations that can be challenging to quantify begin to emerge while zooning the diversity of landscapes and plant varieties.

6.2.2 The uses of IoT in the 4IR and agricultural advancement

The Internet of Things (IoT) refers to the general interconnection between all external items, where smart nodes within the global Internet network resolve the interaction between humans, objects, and humans themselves (Patil *et al.*, 2012). Much technological advancement has been made in recent years in fields including IoT, Big Data, cloud computing, and mobile computing. The world is heading toward notions like smart cities, smart homes, smart jobs, etc. in the modern day. Improper maintenance harmed the crop, leading to significant losses for the farmer. As a result, the idea of "smart farming" was created (Singh *et al.*, 2018). IoT applications span a wide range of industries, including those in transportation, smart agriculture, marketing, supply chain management, healthcare, and infrastructure monitoring, among others (Patil *et al.*, 2012). IoT and modern technologies when used together in the right way can hasten the modernization of the agricultural system. Several issues related to agriculture that often affect farmers and agriculture could be effectively resolved by using smart IoT in agriculture (Lakhwani *et al.*, 2019). The IoT-based CPS provides farmers with a platform to keep track of real-time information and makes significant contributions to the development of smart agriculture. In this, a mobile Android application is utilized to regulate and remotely monitor the crop area, minimizing the need for human intervention (Mekala and Viswanathan, 2017). In order to eliminate inefficiencies and enhance performance across all markets, the IoT is starting to have an impact on a variety of sectors and businesses, ranging from manufacturing, health, communications, and energy to the agriculture industry.

Since the beginning of farming, numerous innovations have emerged to increase productivity and support farmers in resolving problems and crop illnesses. It has been noted that there hasn't been much advancement in the agriculture industry; during the last ten years, the development of a computerized farming system made possible by IoT aids in decision-making, farming, and the early prediction of unfavorable events. Farmers will benefit because it increases crop quality and yield. While automatic discovery is preferable to this drawn-out process of observations by the expert, the automation system of soil component detection, where the result comes from simply observing the change in soil, makes it more affordable and accurate.

Thus, technology aids in early soil condition detection, which helps determine the ideal window for planting seeds and alerts farmers early on. The current learning process focuses on how to integrate IoT with sensor monitoring approaches. The following Figure 6.1 shows the integration of IoT in the agricultural sector.

6.2.3 The context of Bangladesh and Its agriculture in the context of 4IR

Bangladesh is a deltaic nation in South Asia. It has a total area of 147,570 square kilometers (km^2), the eighth-largest population in the world (161 million), and the 13th-highest population density in the world (1,091 people per km^2) (Timsina *et al.*, 2018). In Bangladesh, which has a land area of

Figure 6.1 Key Drivers of Technology in Agriculture.

Source: Ayaz *et al.* (2019)

130,170 km^2, agriculture provides the primary means of subsistence for 70% of the population's livelihood (Yeasmin *et al.*, 2020). In 2018, agricultural land made up 70.7% of Bangladesh's total land area, while arable land's share of total land area decreased gradually from 67.7% in 1969 to 59.7% in 2018 (Knoema, 2018). By 2030 and 2050, respectively, Bangladesh's population will reach 186 and 202 million (Timsina *et al.*, 2018).

Humans have several basic needs, but one of them is food. Food is now a requirement that needs to be given in large quantities and in high quality due to the growing population of the world. Agriculture and livestock are two fields that need to be developed to increase the availability of food (Srisruthi *et al.*, 2016). In order to feed the 9.6 billion people on the planet and the 220 million people in Bangladesh, food production must rise by 70% by 2050, which will provide significant problems for the agricultural sector. Bangladesh has established itself as a global leader in the agriculture industry as a result of its supportive agricultural policies and successful application of contemporary technology. Currently, it ranks second in the production of jute, third in dairy, fourth in rice, fifth in beef, seventh in potato, and tenth in fish. However, despite the fact that modern technologies have penetrated the agricultural industry, a sizable portion of farmers, producers, and growers have yet to benefit from IoT-based smart devices and their applications, which calls for further research. Through Vision 2041, Bangladesh too intends to reach upper middle-income status by 2040. It must look at the possibilities of IoT-based agriculture and the 4IR through agriculture in order to accomplish this goal, along with potential challenges.

Agriculture is one of the oldest industries (Horng *et al.*, 2019). In Bangladesh, agriculture remains the largest sector in terms of employment, providing 50% of the labor force, and about 70% people overall depend on agriculture for their livelihood (Yeasmin *et al.*, 2020). The other indirect contribution of agriculture is to provide basic raw materials for industrial processing. At present, many agro-processing industries are fully dependent on agriculture for basic raw materials, which include rice milling, sugar, tea, fruit juice, spices, edible oil, tobacco, jute textiles, cotton textiles, starch, etc. (Asaduzzaman *et al.*, 2021). In the case of crops, rice output has gone up from around 12 million metric tonnes in the late 1970s to more than 36 million metric tonnes in 2021; aman output has risen from 7–8 million metric tonnes to 13–14 million metric tonnes. Potato is another crop, which saw an increase in output to 10 million metric tonnes in 2021 from about 1 million metric tonnes in the mid-1980s; such an increase is almost ten times higher.

The issue of insufficient rice production has largely been resolved in Bangladesh; however, the country's performance with other crops is at best inconsistent and occasionally disastrous. Official statistics indicate that during the past 20 years or so, fish production has increased at a rate of about 5% annually. Livestock and poultry products are now more readily available, as well as more varied. Commercial poultry farms that produce both meat and eggs have grown significantly lately. According to one estimate, 20 million eggs are produced per day, which is a difference of more than 50% from the 7.34 billion eggs produced annually.

Bangladeshi farmers continue to cultivate their land using traditional techniques, which causes a low output of fruits and vegetables. Using automatic machinery can thereby increase crop productivity. Farmers need innovative, technologically based production techniques to meet these needs so they can produce more with less land and labor. Additionally, the world is moving toward IoT-based concepts for a smarter future, such as smart cities, smart homes, and smart workplaces. This is the rationale behind the introduction of smart farming, not just in Bangladesh but throughout the world. To show off the smart and intelligent capabilities of conventional agriculture, the sector is developing and employing technology found in an IoT system. Such advancements in smart agriculture, which reduce costs, enhance quality, and maximize yields, bring about revolutionary changes in the production, harvesting, storage, and sale of agro-products and significantly contribute to the development of 4IR.

6.2.4 Benefits of smart agriculture

The term "smart agriculture" refers to the collection, processing, analysis, and use of data by autonomous control systems that, when coordinated, maximize farm productivity and profitability (Srisruthi *et al.*, 2016). It is a revolutionary state that collects and represents real-time data on the agriculture production environment and makes it simple for agricultural facilities to access data for cultivation, irrigation, harvesting, marketing, and disaster prevention. It is an automated and directed information technology-based system that was implemented with the IoT (Mekala and Viswanathan, 2017). With the use of sensor devices including light, temperature, relative humidity, and soil moisture sensors, smart agriculture offers the Greenhouse Monitoring System based on IoT with cloud-based capabilities that can effectively monitor various environmental data (Keerthi and Kodandaramaiah, 2015). The remotely sensed thermal inertia technique in smart agriculture is recognized as a smart way for retrieving soil moisture at the ground surface. The land surface soil dampness in dry regions can be roughly calculated using the thermal inertia model, which is based on an approximation of the energy budget equation and the heat conduction equation at the terrestrial surface using the calculated land surface temperature and reflectance from a moderate-resolution imaging spectroradiometer. Additionally, land surface soil dampness was calculated using a thermal inertia soil dampness model and secondary information such as bulk density and soil texture (Ma *et al.*, 2013).

Using ZigBee as the transmission technology, an automated irrigation system is created for smart agriculture to monitor and manage the various factors derived from an agricultural field, such as humidity, water level, temperature, and human interaction, in order to optimize the use of water, electricity consumption, and labor costs (Khelifa *et al.*, 2015). With a drip irrigation system, water waste can be reduced, and the system operates using data from water-level sensors (Mekala and Viswanathan, 2017). Herbicides are sprayed consistently over the entire field in traditional weed control methods. This method wastes a lot of resources, contaminates the environment, and causes health issues in people because only 20% of the spray actually reaches the plant, and less than 1% of the chemical actually helps to manage weeds (Ghazali *et al.*, 2008). A prototype of the automatic weed detection and a smart weed control system should be used in order to detect weeds automatically among plantations in order to prevent these problems. These systems must be able to identify weeds in the field so that pesticide sprayers can be instructed to apply their chemicals directly where they are needed. Additionally, it emphasizes minimizing the use of pesticides that hinder plants' natural growth as well as expensive labor (Aravind *et al.*, 2015).

The use of robotics-based guiding systems in smart agriculture is particularly beneficial for agricultural purposes where it is necessary to spray pesticide on various crops. This robotic pesticide sprayer can go through fields and has sensors to find vegetation on both sides. Farmers utilize a tiny tank for the motor and insecticides in this technique. It will automatically begin to spray if a plant is detected. Additionally, this device contains a wireless camera that can take pictures and show them on the display. Agricultural products need security and protection from the very beginning, such as defense against rodent or insect attacks in fields or grain storage, and not just in terms of resources. In order to address issues like identifying rodents, crop hazards, and providing real-time notifications based on information analysis and processing without human interaction, IoT-based devices can be integrated into agricultural fields, grain warehouses, and cold stores for security purposes (Baranwal *et al.*, 2016).

IoT technology in smart agriculture creates a remote crop harvesting system that uses technology and clever picture recognition to assess crop maturity. Robotic arm harvesting is a semi-automated technique that cuts down on the time farmers must spend on monotonous work. Users can eventually decide when to harvest crops thanks to the system's transmission of findings showing whether crops are adequately ready for harvesting to a webpage. The user webpage allows farmers to control their harvest remotely and shows various data that affects harvesting choices, such as streamed images, post-detection results, the distance determined by ultrasonic sensors, arm control schematics, arm control buttons, and the GPS position of tracked mobile vehicles (Horng *et al.*, 2019). The other choices for agricultural infrastructure are largely missing here without a proper marketing framework. Farmers can negotiate with middlemen for a fair price and make lucrative judgments about what to produce and what price to expect, utilizing the fast and objective agricultural marketing supported by IoT-based information systems (Kundu *et al.*, 2018).

6.3 METHODOLOGY

This study employs a qualitative approach to obtain the set objectives. Understanding human ideas, behavior, attitudes, and perceptions of particular social or non-social issues within their own contextual circumstances is the goal of qualitative research. The open-ended questions used in qualitative research methods encourage participants to freely express their opinions and beliefs. The goal of qualitative research is to comprehend not just "what" but also "why" individuals believe the way they do. In their responses, participants describe the reasoning behind their choices and beliefs, and qualitative research techniques analyze the resulting data to produce generalizable findings for a broader group of comparable types. Social sciences like sociology, anthropology, and psychology have a significant influence on qualitative research. As a result, it enables the researcher to delve deeply into the data collection process and look for as much qualitative data as they can, without restrictions on the questions they ask and the responses they receive from the participants.

6.3.1 Data collection and data analysis

The study collected primary data using in-depth interview technique from 12 respondents. The respondents of the study are pertinent to the study's objectives. All the respondents are directly or indirectly related to smart agriculture in Bangladesh. Some of the respondents have been providing all sorts of facilities to promote smart agriculture in Bangladesh, and some other respondents are implementing smart agriculture in the ground. The information gathered from the respondents is rich in quality due to their deep knowledge in the respective fields. Moreover, the farmers who have been farming agricultural products using sophisticated technologies shared their experiences about technological implications in the agriculture in Bangladesh.

This study collected seven face-to-face interviews with the informants, and the rest five interviews were conducted online using the famous online meeting platform Zoom. The sample (respondents) of this study was selected purposefully so that the researcher can gather the required information which is relevant to the study topic. The respondents of the study were asked open-ended, semi-structured questions to gather relevant information about the existing smart agriculture's prospects and challenges in Bangladesh. All the interviews lasted for maximum 45 minutes. The interviews were recorded with the permission of the respondents. Note taking technique was also applied where there was an obligation to record the interview session. All the protocols have been maintained carefully before starting the interview with the respondents.

The collected interview data has been analyzed thematically. The themes of this study have been obtained systematically. For example, the interviews were recorded using an audio recorder, and those conducted online were also recorded. However, note-taking techniques were also applied where it was an obligation to record the interview session. Subsequently, the recorded interviews were transcribed from the beginning to the end. The transcriptions were carefully read. Then the important issues or keywords were coded separately. After receiving the unique codes, the researchers prepared some subcategories using the codes of the interview data. Later, the sub-categories were converted into categories and finally into themes. The study obtained three unique themes from the interview data. The data analysis chronology is given in Figure 6.2.

6.3.1.1 Demographic discussion

The study has considered 12 respondents from diversified professions, educations, ages, gender, and experiences. The respondents are classified according to their professional background. For example, three of the respondents were farmers who are directly related to the application of smart agriculture in their lands. Out of 12 respondents, 5 respondents were agricultural officers from the Bangladesh Agricultural Development Corporation (BADC), the Department of Agricultural Extension (DAE), Bangladesh, the Bangladesh Agricultural Research Institute (BARI), and the Bangladesh Rice Research Institute (BRRI). Furthermore, three of the respondents were professors from

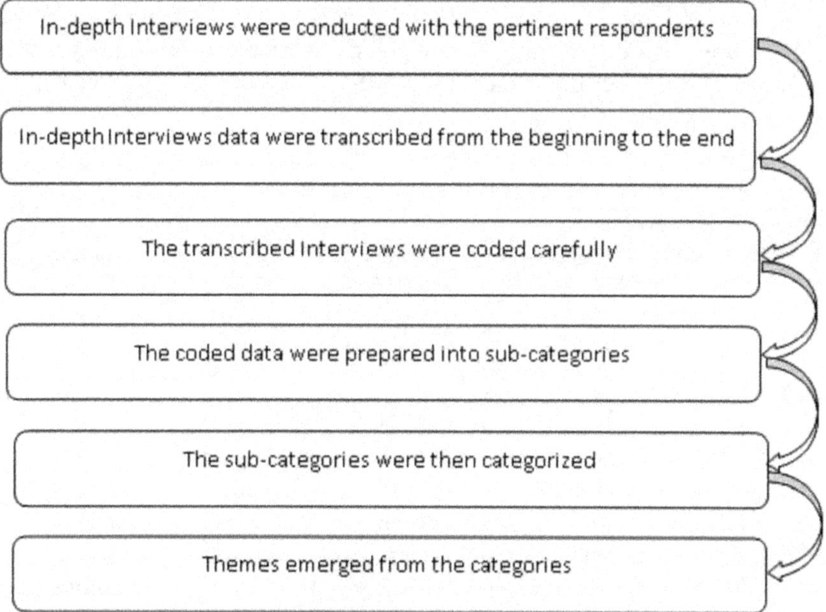

Figure 6.2 Chronology of Interview Data Analysis/Thematic Data Analysis.

three reputed universities, namely the Bangladesh Agriculture University, the University of Rajshahi, and Hajee Mohammad Danesh Science and Technology University. Apart from these, the respondents are also diversified in accordance with their ages, educations, and number of years of experience. Table 6.1 indicates the demographic information for this study.

6.4 DISCUSSION OF THE EMERGED THEMES

The study emerges with three unique themes. These themes comply with the set objectives of the study. The climate of Bangladesh is diversified, and farmers grow several grains in different seasons. However, in recent years, technological advancement and mechanization of the agricultural sector have dramatically changed agricultural production in Bangladesh. The three themes are shown in Figure 6.3.

6.4.1 Smart agriculture and its prospect in Bangladesh

The Fourth Industrial Revolution has brought about revolutionary changes in the fields of industry, agriculture, and services. Smart and novel technologies have been rapidly utilized in the respective sectors. The field of smart agriculture was developed in response to the need to improve agricultural productivity in order to address challenges with food security and environmental impact. In terms of productivity and sustainability, precision agriculture (or smart farming) can greatly increase agricultural production. Despite the fact that productivity seems to be the driving force behind

Table 6.1 Demographic Profile of the Respondents

Sl. no.	*Respondents code*	*Gender*	*Age*	*Profession*	*Experience*	*Education*
1	FAR 1	M	52	Farmer	30 Years	Under HSC
2	FAR 2	M	48	Farmer	25 Years	SSC
3	FAR 3	M	49	Farmer	29 Years	Under SSC
4	BADCO1	M	45	Officer	15 Years	Post-Graduation
5	BADCO2	F	40	Officer	9 Years	Graduation
6	BRRI1	M	52	Officer	23 Years	Doctorate
7	BAEO1	M	38	Officer	10 Years	Graduation
8	BAEO2	F	35	Officer	6 Years	Graduation
9	BAUP1	M	48	Professor	18 Years	Doctorate
10	RUP1	F	50	Professor	22 Years	Doctorate
11	RUP2	M	52	Professor	24 Years	Doctorate
12	EP1	M	35	Entrepreneur	10 Years	Graduation

Source: Authors

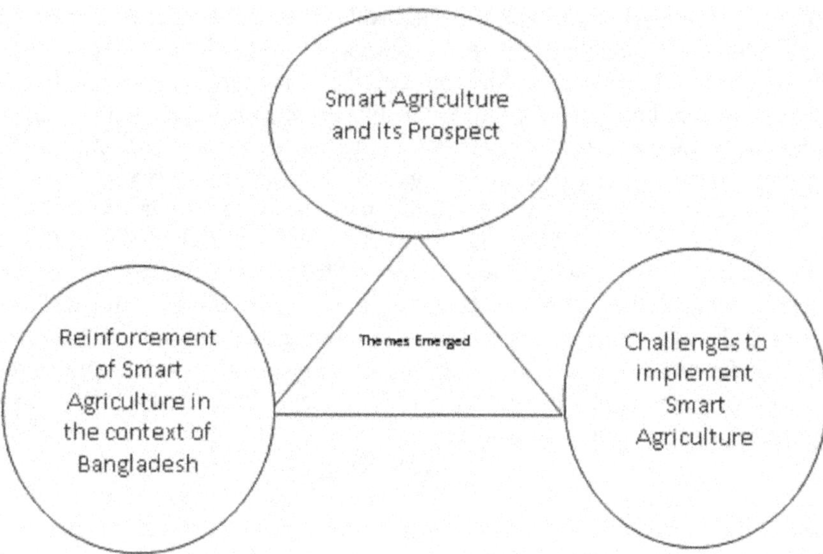

Figure 6.3 Emerged Themes.

every technological advancement in agriculture, sustainability should not be undervalued. The reduction of the environmental impact of agriculture activities is one of the goals of smart agriculture, since sustainability appears to be a significant concern across the spectrum of human activity.

6.4.1.1 Development of climate smart agriculture

A strategy for changing and reorienting agricultural systems to enhance food security in light of the new climate change realities is known as climate-smart agriculture (CSA). The majority of the world's poor, who depend on agriculture for their livelihoods, are made more vulnerable by widespread changes in rainfall and temperature patterns that endanger agricultural production. Market disruptions caused by climate change put the availability of food at risk for the entire population. By enhancing farmers' capacity for adaptation as well as agricultural production systems' resilience and resource use effectiveness, threats can be lessened. CSA encourages coordinated efforts by farmers, researchers, the commercial sector, civic society, and decision-makers to forge routes that are climate-resilient.

> For the purpose of decreasing poverty and ensuring food security, as well as in the larger context of urbanization and the expansion of the non-farm sector, it is essential to increase agricultural productivity and incomes in the smallholder production sector.
>
> #RUP3; Age: 55; Professor; Experience: 24 years; Doctorate Degree; Date: 20/10/2022

To meet rising demand, agricultural production will need to increase by 60% globally by 2050. The majority of this growth will need to come from higher productivity (United Nations, 2009). Agriculture's growth is already being hampered by climate change. The Intergovernmental Panel on Climate Change (IPCC) asserts that climate change has an impact on food output in many parts of the world, with negative consequences outweighing good ones increasingly frequently. Developing nations are particularly sensitive to further negative effects. According to projections, climate change has already decreased worldwide wheat and maize yields by 3.8% and 5.5%, respectively (Tripathi *et al.*, 2019). Several researchers also foresee severe declines in agricultural output when temperatures cross crucial physiological thresholds (Elder and Hayashi, 2018). The ability of farmers to adapt is put to the test by increased climate variability (World Agriculture, 2019). Both rural and urban populations face threats to their ability to get food as a result of climate change's decreased agricultural revenue and production, increased hazards, and market disruption. Poor farmers, those without access to land, and marginalized ethnic groups are especially at risk (Hassan, 2018).

The main goal of CSA is to support initiatives at all scales—local, national, and international—to sustainably use agricultural systems to ensure food and nutrition security for all people at all times while including essential adaptation and capturing possible mitigation. To accomplish this goal, three goals are specified: (United Nations, 2017) sustainably raising agricultural productivity to support equitable income growth, food security, and development; preparing for and enhancing resilience to climate change at all scales, from the farm to the national level; and (United Nations, 2018) creating opportunities to lower GHG emissions from agriculture than in the past (Elijah *et al.*, 2018). The following verbatim addresses the situation of the climate change issue in Bangladesh.

> The climate has changed rapidly, and it has severe impact on the agricultural yield. It is now time to adjust agriculture with the climatic conditions. The use of mechanism, sensors, internet of things and other cutting-edge technologies have been gaining popularity globally in these days. Although maximum number of farmers in Bangladesh is still applying conventional method of farming, it is expected that soon more and more farmers will use smart technologies in the agricultural production.
>
> #BADCO2; Age: 40; Female; Officer; Experience: 9 years; Graduation; Date: 15/10/2022

6.4.1.2 *The prime goal of smart agriculture and the issue of food security*

The UN and the world community set a goal to eradicate hunger by the year 2030 in a plan referred to as "The 2030 Agenda for Sustainable

Development," which was unveiled in 2015. However, recent data from the WHO (World Health Organization) do not appear to be sufficiently positive to support the agenda, as one in nine people worldwide—more than 800 million people—are experiencing a food shortage (WHO, 2018). Even if these figures on their own are fairly disturbing, the quality of the food is even more distressing. In addition to accessibility, food quality is growing to be an even bigger problem.

The demand for food is rising as a result of changing diets and a growing worldwide population. As food yields plateau in many regions of the world, ocean health deteriorates, and natural resources—including soils, water, and biodiversity—are severely depleted, and production is straining to keep up. According to a 2020 estimate, the number of hungry people has increased by about 60 million over the last five years to reach close to 690 million, or 8.9% of the world's population. As the globe must produce almost 70% more food by 2050 in order to feed an expected 9 billion people, the problem of food security will only get worse. Major features offered by IoT-based systems include data collection and communication infrastructure, cloud-based intelligent data analysis, decision-making, end-user interfaces, and operation automation (used to connect smart items to end-user applications through the Internet). The agricultural industry is expanding in new directions thanks to these capabilities.

> I must say that smart agriculture positively influences food security at the individual as well as national level. Today, the world has been facing with several problems. Weather adversity and war and geo-political issues significantly poses threat to the supply of food worldwide.
>
> #BAEO1; Male; Age: 38 years; Experience: 10 years; Graduation; Time: 13/10/2022

6.4.1.3 Greenhouse farming

Precision farming and greenhouse farming are somewhat comparable, although with slightly different goals. The primary distinction is that greenhouse farming is carried out in a confined or isolated area where environmental variables are regulated and managed by sophisticated technologies. Despite the fact that greenhouse farming is not a new practice, information technology and the Internet of Things (IoT) have discovered uses that are compatible with greenhouse practices like controlling temperature and humidity as well as monitoring in the shed. Without IoT and smart systems, precise and ongoing monitoring and control are impossible. Although greenhouse farming produces higher output than conventional methods, its area is relatively smaller than that of open-field farming. One of the smaller nations and the second-largest exporter of agricultural products using hydroponic and greenhouse farming methods is the Netherlands

(Villarrubia *et al.*, 2017). We can also employ desert territory for sustainable cultivation through the use of greenhouses (Neto *et al.*, 2014).

> Previously greenhouse technology was not that much famous. People did not know the benefits of this technology, but this technology is very much effective when it comes to grow crops and vegetables round the year regardless of its seasonal impact. Moreover, harmful insects cannot harm the plants inside the greenhouse farms. Moreover, these plastics are basically ultraviolet ray protected.
>
> #EP1, Age 35 years, Experience: 10 years; Education: graduate; 20/10/2022

6.4.1.4 Urban farming

Urban farming is a relatively new idea that combines several cutting-edge techniques, including rooftop, indoor, vertical, hydroponic, aquaponic, and aeroponic farming. As was already said in the chapter, more people are moving into cities, and these cities are the biggest consumers of food. On the other hand, climate change and water shortages have a negative impact on agriculture and present significant difficulties for long-term food security. The considerable distance between food producers and consumers creates transportation and chain supply costs that have an impact on food quality and add to pollution from transport vehicles. People can grow food nearby through urban farming as a solution to their need for affordable and fresh food. Urban farming depends entirely on a precise control environment and system that allow it to flourish day and night, all year long, regardless of the seasons or weather.

> In the mega city like Dhaka where vacated place is very much scarce, yet people are growing plants in their house rooftop with the help of smart agricultural knowledge. Some people are doing such farming out of passion and hobby, and some are doing it professionally.
>
> #BADCO1; Age 45 years; Officer; Experience: 15; Education: Postgraduate; 18/10/2022

6.4.1.5 Smart irrigation

It is possible to develop weather-adaptive smart irrigation systems and use less of the limited water supply by utilizing IoT and smart systems. Depending on the crop and soil stress levels, smart irrigation is intended to only irrigate when absolutely essential. Sprinkler irrigation can be done utilizing UAV for precise on-the-spot irrigation in order to deal with the variable aspect of water stress.

> This is one of the important advancements in the agricultural system. Previously, farmers used to irrigate more water in their farmlands which was absolutely unnecessary. However, in the recent years, we have been able to make farmers understood that smart irrigation system

automatically measures how much water the plant actually requires and thus wastage and cost have significantly been minimized.

#BAEO2, Female; Age 35 years; Officer; Experience: 6 years; Education: Graduate; 15/10/2022

6.4.1.6 *Pest and weed management system*

The crop can be severely impacted by pests, weeds, and pathogens, and simply weeds can lower yield by up to 30% [61]. However, pesticides and herbicides can have a negative impact on profits and product quality, which is a major worry for consumers. Smart vehicles can also be used for this purpose. IoT and smart systems are able to identify disease, pests, and weeds in the crop at an early stage and alert the farmer. They are also capable of eliminating pests and diseases by precisely targeting them with pesticides and herbicides.

> The excessive use of chemical pesticide is always discouraged. Now-a-days, smart technology can estimate the health of the plants and accordingly prescribe for required actions. Instead of chemical fertilizer we often suggest providing organic fertilizer.
>
> #BRRI1, Age 52 years, Officer; Experience: 23 years; Education: Doctorate, 19/10/2022(via Zoom)

6.4.2 Challenges to implementing smart agriculture

There are a number of challenges to implementing smart agriculture in Bangladesh. The following challenges mentioned in Figure 6.4 have been highlighted. These challenges are the main building blocks for implementing smart agriculture in Bangladesh. Among these challenges, *knowledge gaps* regarding smart agriculture impede the promotion of smart agriculture significantly. In Bangladesh, most of the farmers are uneducated or unaware about the application of cutting-edge technologies in agricultural activities.

> The implementation of smart agriculture in Bangladesh is very difficult task due to the knowledge gap of farmers about smart technologies. Many farmers think that this is probably a complicated task to incorporate technologies in their agricultural activities. Moreover, they do not often possess enough knowledge to operate modern machines.
>
> #BAUP1, Age 48 years, Professor; 18 years; Education: Doctorate; 20/10/2022

> Truly speaking, I don't know much about smart agriculture. Whenever we faced any problem we resorted to the agricultural department, they often help us with their equipment. We need more training on these agricultural systems.
>
> #FAR 3; Age 49; Profession: Farmer; Experience: 29 years; Education: Under SSC; 09/10/2022

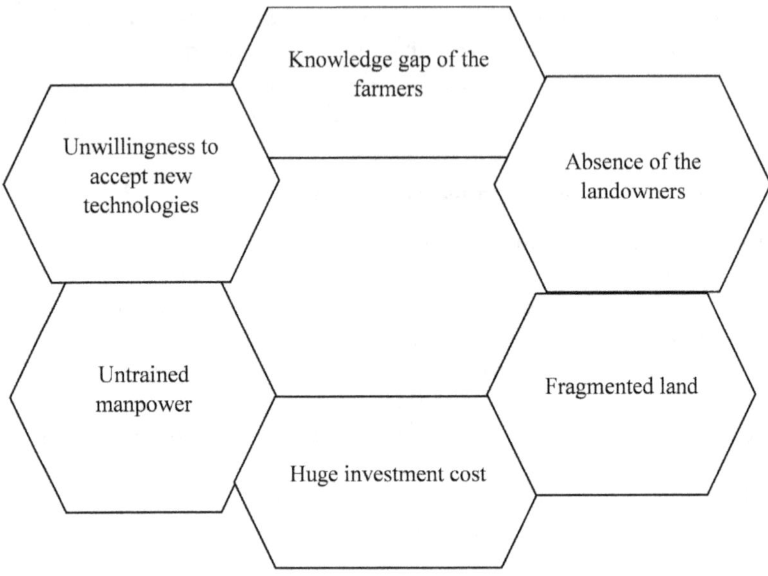

Figure 6.4 Challenges of Smart Agriculture.

In Bangladesh, the *landlords are often absent* from agricultural activities. The owners of the land remain engaged in some other activities other than agriculture. They often employ people to cultivate their land contractually. Besides, this challenge, *fragmentation of cultivable land* is another prominent issue that restricts the implementation of smart agriculture. The land in Bangladesh is divided into small portions due to the separation of several owners. As a result, a large chunk of land is required to implement smart agriculture effectively. On the other hand, different landowners have different opinions regarding smart agriculture.

> It is difficult to implement smart agriculture in Bangladesh as the owner of the land is absent from their land, they are sometimes not aware of any cultivation in their land. Even sometimes, the landowner does not know which crop is going to be ploughed by his nominated contractual farmers. This is how it is very difficult for the farmers to cultivate the lands using advanced technologies.
>
> Moreover, fragmented or division of land is another problem here. Due to having many rightful owner of a small chunk of land, the agricultural production becomes interrupted, besides, it is difficult to apply mechanical equipment like trucker and other advanced technologies due to a very small fraction of land.
>
> #BAUP1, Age 48 years, Professor; 18 years; Education: Doctorate; 20/10/2022

The *initial investment cost* of technological integration is pretty high to implement smart agriculture in Bangladesh. The farmers in Bangladesh are often poor, and they barely manage the required equipment to cultivate their small lands. As the technological installation costs are pretty high, farmers cannot afford such new technologies to implement smart agriculture on their farmlands. Consequently, they prefer conventional ways of cultivating their lands.

> We are having a very hard time to motivate farmers to apply smart agriculture because it is pretty expensive for poor farmer to afford such an expensive equipment and technologies at the very beginning of their cultivation. However, if we could arrange these huge initial investments available at the farmers door, they would happily introduce smart agriculture in their cultivable land.
>
> # BAEO1; Male; Age: 38 years; Experience: 10 years; Graduation; 13/10/2022

The manpower in the agricultural development and extension departments *is not well trained* practically, which is a barrier to implementing smart agriculture. Moreover, the farmers usually do not get proper training to apply technological advancement to the farmland and end up with a conventional agricultural system.

> It is indeed a challenging task for us as we do not have sufficient number of trained officers who could give hands on training to the farmers and new entrepreneurs to apply smart technologies in the agricultural sector.
>
> # BRRI1, Age 52 years, Officer; Experience: 23 years; Education: Doctorate, 19/10/2022 (via Zoom)

Unwillingness to accept new technologies is another challenge to developing smart agriculture in Bangladesh. The farmers in Bangladesh have been applying the conventional agricultural system for a long time, and they have gained vast expertise in this regard. Hence, they do not show willingness to accept news and smart technologies in agricultural activities in Bangladesh.

> The farmers have unwillingness to adopt smart agriculture due to a number of reasons, for example, knowledge gap, high costs, availability of sophisticated technologies around them, ability to apply advanced knowledge, proper training, ICT use and data analysis and many more.
>
> #HSTU2; Female; Age: 46; Professor; Experience: 15 years; Education: Doctorate; 24/10/2022 (via Zoom)

6.4.2.1 *Use of the internet and sophisticated ict by most of the uneducated farmers*

The outreach of the internet is one of the key issues when it comes to promoting smart agriculture in remote areas in Bangladesh. Moreover, the use of the internet by the marginal farmers in the villages is also a key matter to take into consideration here. The poor farmers often cannot afford to purchase smart mobile phones connected to the internet and use information and communication technologies. Moreover, there remains a huge knowledge gap regarding the application of ICT, especially to implement smart agriculture in the remote areas in Bangladesh.

> We are living at the age of 4IR and smart agriculture and using cutting edge devices to catch up with the fast-moving world. However, our farmers in the remote areas are still lacking required knowledge to cultivate their small chunk of farmland in a smart way. I think wide array of arrangement is required to educate these poor farmers to use internet and ICT by providing them smart phones otherwise the dream to fulfill smart agriculture in the industrial revolution will simply go in vain.
>
> # BAUP1; Age 48 years; Professor; Experience: 18 years; Education: Doctorate; 20/10/2022

6.4.3 Reinforcement of smart agriculture in Bangladesh

The implementation and success of an approach are impossible if there is no *proper coordination* among the several departments. The government agencies are independent to accomplish their objectives, and one government agency often does not care about another department. However, the farmers at the grassroot level have to face the hassle of getting services regarding the smart agriculture. Therefore, there should be a proper coordination among the several government agencies that are working to implement smart agriculture in Bangladesh. For example, the Bangladesh rural agricultural department, the Bangladesh agricultural expansion department, and the information and communication technology department should maintain coordination to successfully implement smart agriculture in Bangladesh.

> No way can a vision see the face of success if there lacks coordination among the departments. Tell me, who is paying the price for such disorganization of the bodies. Yes, the farmers. Hence, the government should take action to make a chain of the departments and the aim of these departments should only be citizen charter.
>
> #RUP3; Age: 55; Professor; Experience: 24 years; Doctorate Degree; 20/10/2022

The big data generated from industrial and agricultural sensors should be stored, processed, and retrieved from a cloud server. Information generated from big data is very much essential to boost smart agriculture in Bangladesh. Farmers, academics, researchers, and development agencies should be given access to the data sources so that they can further continue with decision-making, research, and funding activities to support smart agriculture in Bangladesh. Figure 6.5 explains such gateway services.

Contrary to earlier farming, the majority of duties in modern, large-scale agriculture are carried out by *powerful, urbane machinery* like tractors, harvesters, and other robots that are either fully or partially supported by remote sensing and other communication technology. In precision agriculture, the operational vehicles are outfitted with GPS and GIS capabilities so they can work precisely, site-specifically, and autonomously when operations like sowing, fertilizing, watering, and harvesting are being conducted. Over the past few decades, agriculture has changed from small- to medium-sized farms to highly industrialized and commercial farming. This change enables the major corporations to treat agricultural similarly to other sectors, such as manufacturing, where measurements, data, and control are crucial for

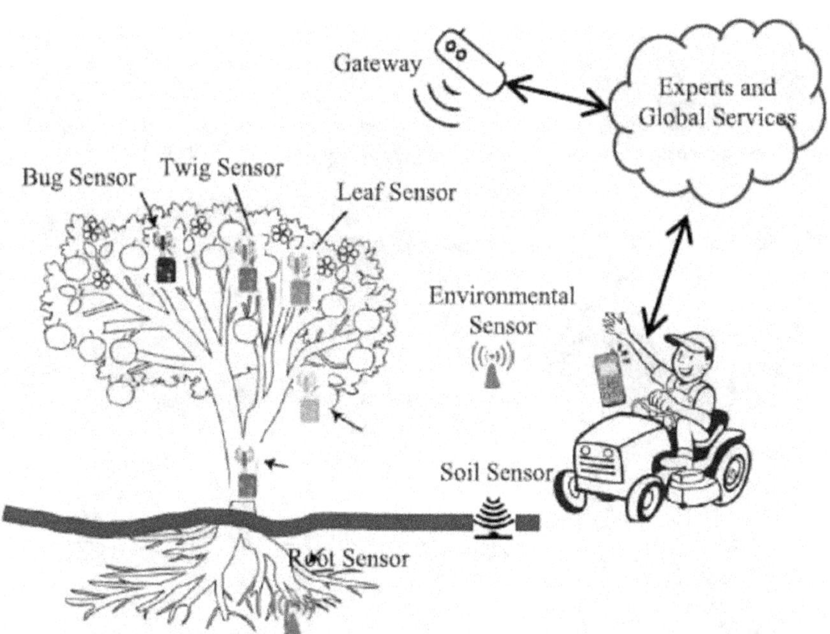

Figure 6.5 Big Data Gateway Services.

Source: M. Ayaz *et al.* (2019)

maintaining a balance between costs and output in order to increase profits. IoT technologies and solutions will thereby benefit every area of agriculture that can be automated, digitally planned, and managed. Due to this reality, attempts are being made to provide more advanced tools, such as agricultural robots, to carry out a variety of tasks, including planting, watering, weeding, picking, thinning, fertilizing, spraying, packing, and transportation. This revolution is being fueled not only by technological improvement but also by the demand for better and more affordable food, as well as other causes, including concerns over losing cheap labor. Figure 6.6 shows the incorporation of tools and mechanism in the agricultural sector to implement smart agriculture.

Meanwhile, the government should establish wireless connectivity infrastructure for *end-to-end communication infrastructure*. In order to deliver the services required for smart agriculture, the end-to-end communication possibilities can be segmented into multiple layers that interact with one another. Due to the fact that this wireless infrastructure relies on narrowband or ultra-narrowband technologies, the data is encoded using very small spectrum fragments by altering the carrier radio wave's phase. These features allow it to work at a high level even when 100 sensors are required

Figure 6.6 Selected IoT-Based Products and Prototypes for Smart Agriculture.
Source: Ayaz *et al.* (2019)

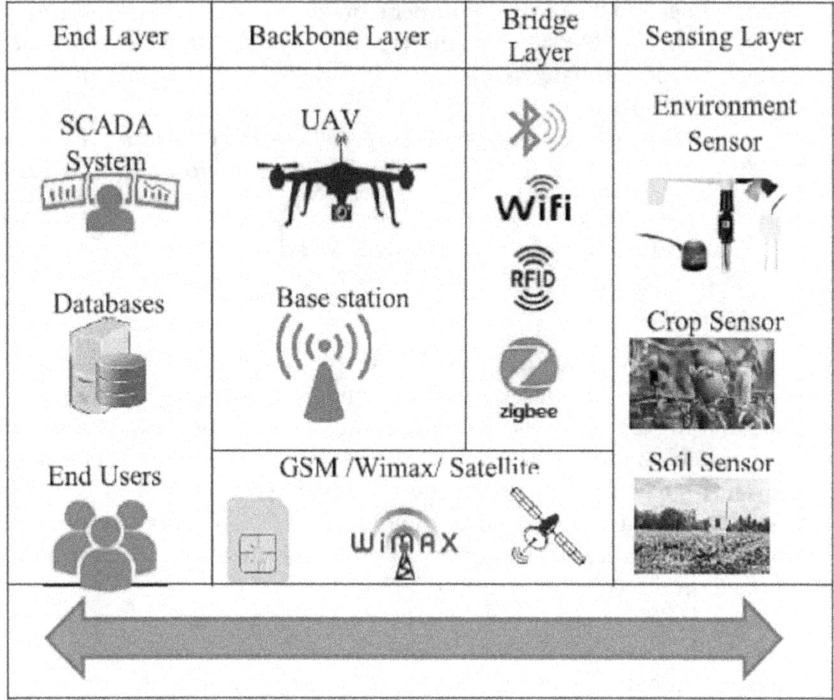

Figure 6.7 End-to-end Communication for Smart Farming.

to relay data simultaneously. Figure 6.7 shows the end-to-end networking communication infrastructure at different levels.

6.4.4 Cloud-based storage, utilization, and analysis of big data efficiently

In this modern era of technology, data is a big asset. Data-generated information will help someone understand and estimate what is required for him to maneuver resources efficiently. The sensors installed in the farmland will measure irrigation, pesticides, fertilization, and weather forecasts. These data will be generated every now and then. Hence, it is key for the farmers to get access to these data so that they can take precautions and proper actions when it comes to crop development. These big data require a cloud-based storage system and uninterrupted access to them by the key persons so that they can download them any time they want and analyze to get maximum yield.

Everybody should talk about data because data can give you surprising results to implement smart agriculture. But before jumping into the data we need to make sure a proper storage and off course cloud based, so that experts like us can analyze and produce empirical evidence which would ultimately help a greater extension of agricultural sector of Bangladesh.

#BAUP1, Age 48 years, Professor; 18 Years; Education: Doctorate; 20/10/2022

6.4.5 Compulsory smart agricultural education at preliminary, intermediary, and tertiary education

The context of the Fourth Industrial Revolution and smart agriculture should be made a compulsory education among the students at the school, college, and university levels. When the students possess knowledge about making agriculture using smart technologies, they will be highly motivated to introduce smart agriculture beyond just looking for a job. This will also help students change their mindset to not wait for a conventional job after completing their education but rather to try to become entrepreneurs in this vast agricultural sector.

We need to involve more and more educated young generation in this sector (agriculture) and there should be proper infrastructure to give entrance to these young people in to smart agriculture. Because, we have to sustain our agriculture in the long run.

#FAR 1; Age: 52; Profession: Farmer; Experience: 30 years; Education: Under HSC; 08/10/2022

6.4.6 A friendly nexus between agriculture and industry

The industrial sector in Bangladesh requires huge raw materials which can be procured from the agricultural sector. For example, the garments sector which is one of the leading export earning sectors in Bangladesh requires huge volume of yarn to produce fabrics. The government has to deplete a big chunk of its foreign reserves only to import these raw materials in different industrial sectors. Hence, there remains a huge opportunity to promote smart agriculture in Bangladesh to produce these needed raw materials for industrial use and processing. This will certainly preserve the foreign reserves, and at the same time, these can be exported to other countries.

Industry is not and never separated from agriculture. Yes, this is true that our agriculture falls short to supply the requirements of the industry, but I think, if smart agriculture is given importance and if we can utilize our existing resources smartly, the day is not that far, and we will be one of the leady economies in the world. However, we just need the

> smart people who can use these smart technologies. The further development of Bangladesh is in the hands of smart revolution.
>
> #RUP3; Age: 55; Professor; Experience: 24 years; Doctorate Degree; 20/10/2022

6.5 CONCLUSION

To meet the rising food demand of the growing global population in the face of ever-dwindling arable land, the focus must be on smarter, better, and more efficient crop-producing practices. The adoption of farming as a profession by tech-savvy, creative young people, agriculture as a means of achieving fossil fuel independence, tracking crop growth, safety and nutrition labeling, and collaborations between growers, suppliers, retailers, and buyers are all examples of the development of new methods to improve crop yield and handling. This research chapter endeavors to investigate the prospects and challenges of smart agriculture in the context of the Fourth Industrial Revolution in Bangladesh.

This research work finds that there has been a growing demand for smart agriculture in Bangladesh, especially among the young generation. Many of the unemployed people are engaging more in agricultural projects, and they have been using technologies frequently. The concerned authorities and related departments are playing a significant role in promoting smart agriculture in Bangladesh. The government's focus on food security in Bangladesh has also been highlighted in recent years, and it is expected that the promotion of smart agriculture has the ability to potentially contribute to the vision of the government's food security. However, there remain some challenges like knowledge gap, landlords are often absent, fragmentation of cultivable land, initial investment cost of technological integration in agricultural activities, manpower in the agricultural development and extension departments is not well trained, the unwillingness to accept new technologies on the ground as well, which have been discussed with some empirical evidence. These challenges need to be addressed efficiently through a consolidated effort by all the concerned bodies to promote smart agriculture effectively. Moreover, universities and research institutions should be given opportunities with sufficient funding facilities to carry out research on the issues of smart agriculture and industrial revolutions.

The findings of this chapter are very important for the concerned authorities to formulate policies for the implementation of smart agriculture in Bangladesh. This chapter will also significantly contribute to the existing literature, and students, policy makers, scholars, academics, and researchers will be able to use this piece of research as a reference in future. This research work has to solely depend on the primary data due to the scarcity of secondary data. However, it is expected that a mixed-methods approach would supply more rich information about the research objectives.

REFERENCES

Aravind, L., Zhang, D., de Souza, R.F., Anand, S. and Iyer, L.M. 2015. The natural history of ADP-ribosyltransferases and the ADP-ribosylation system. *Current Topics in Microbiology and Immunology*, *384*, pp. 3–32. https://doi.org/10.1007/82_2014_414. PMID: 25027823; PMCID: PMC6126934.

Asaduzzaman, M., Mahomud, M.S. and Haque, M.E. 2021. Heat-induced interaction of milk proteins: Impact on yoghurt structure. *International Journal of Food Science*, *22*, 5569917. https://doi.org/10.1155/2021/5569917. PMID: 34604378; PMCID: PMC8483934.

Ayaz, M., Ammad-Uddin, M., Sharif, Z., Mansour, A. and Aggoune, E.H.M. 2019. Internet-of-Things (IoT)-based smart agriculture: Toward making the fields talk. *IEEE Access*, *7*, pp. 129551–129583.

Baker, S.B., Xiang, W. and Atkinson, I. 2017. Internet of things for smart healthcare: Technologies, challenges, and opportunities. *IEEE Access*, *5*, pp. 26521–26544.

Baranwal, T., Nitika and Pateriya, P.K. 2016. Development of IoT based smart security and monitoring devices for agriculture. In *2016 6th International Conference—Cloud System and Big Data Engineering (Confluence)*, CFP1669Y-ART, Springer, New York (pp. 597–602). https://doi.org/10.1109/CONFLUENCE.2016.7508189.

Dobele, M. and Zvirbule, A. 2019. The concept of urban agriculture. In *14th International Scientific Conference Students on Their Way to Science* (undergraduate, graduate, post-graduate students) Collection of Abstracts Apr. 26, 2019 (p. 106).

Elder, M. and Hayashi, S. 2018. A regional perspective on biofuels in Asia. *Biofuels and Sustainability: Holistic Perspectives for Policy-Making*, pp. 223–246.

Elijah, O., Rahman, T.A., Orikumhi, I., Leow, C.Y. and Hindia, M.N. 2018. An overview of Internet of Things (IoT) and data analytics in agriculture: Benefits and challenges. *IEEE Internet of Things Journal*, *5*(5), pp. 3758–3773.

FAO. 2019. How to Feed the World in 2050 by FAO. Accessed: Sep. 6, 2019. [Online]. Available: https://www.fao.org/wsfs/forum2050/wsfs-forum/en/

Ghazali, M.F., Yee, O.A. and Muhammad, M.Z. 2008. Do producer prices cause consumer prices? Some empirical evidence. *International Journal of Business and Management*, *3*(11), pp. 78–82.

GlobeNewswire. 2018. Global Smart Farming Market to Reach $23.14 Billion by 2022. Accessed: Oct. 25, 2022. [Online]. Available: https://www.globenewswire.com/news-release/2018/08/02/1546021/0/en/Global-Smart-Farming-Market-to-Reach-23-14-Billion-by-2022.html

Hassan, Q.F. ed. 2018. *Internet of Things A to Z: Technologies and Applications*. John Wiley & Sons.

IPC Report. 2022. UNICEF. Assessed: Dec. 4, 2023. Available: https://www.unicef.org/about-unicef

Kapoor, A., Bhat, S.I., Shidnal, S. and Mehra, A. 2016. Implementation of IoT (internet of things) and image processing in smart agriculture. In *2016 International Conference on Computation System and Information Technology for Sustainable Solutions (CSITSS)* (pp. 21–26). IEEE.

Keerthi, V. and Kodandaramaiah, G. N. (2015). Cloud IoT based greenhouse monitoring system. *International Journal of Engineering Research and Applications*, *5*(10), pp. 35–41.

Khelifa, A., Touafek, K., Moussa, H.B., Tabet, I. and Haloui, H. (2015). Analysis of a hybrid solar collector photovoltaic thermal (PVT). *Energy Procedia*, *74*, pp. 835–843.

Knoema. 2018. Thyolo—Crude Birth Rate, 1918–2018. Available: https://knoema.com//atlas/Malawi/Thyolo/Crude-birth-rate

Kopnina, H. and Washington, H. 2020. Conservation and justice the anthropocene: Definitions and debates. In *Conservation: Integrating Social and Ecological Justice*, Springer, New York (pp. 3–15).

Kundu, A., Denis, D.M., Patel, N.R. and Dutta, D. 2018. A geo-spatial study for analysing temporal responses of NDVI to rainfall. *Singapore Journal of Tropical Geography*, *39*(1), pp. 107–116.

Lakhwani, K., Gianey, H., Agarwal, N. and Gupta, S. 2019. Development of IoT for smart agriculture a review. In *Emerging Trends in Expert Applications and Security: Proceedings of ICETEAS 2018* (pp. 425–432). Springer.

Ma, M., Guo, L., Anderson, D. G. and Langer, R. 2013. Bio-inspired polymer composite actuator and generator driven by water gradients. *Science*, *339*(6116), pp. 186–189.

Mekala, M.S. and Viswanathan, P. 2017. A survey: Smart agriculture IoT with cloud computing. In *2017 International Conference on Microelectronic Devices, Circuits and Systems (ICMDCS)* (pp. 1–7). IEEE.

Milman, O. 2015. Earth has lost a third of arable land in past 40 years, scientists say. *The Guardian*, *2*, p. 12.

Neto, A.J.S., Zolnier, S. and de Carvalho Lopes, D. 2014. Development and evaluation of an automated system for fertigation control in soilless tomato production. *Computers and Electronics in Agriculture*, *103*, pp. 17–25.

Patil, P.N., Sawant, D.V. and Deshmukh, R.N. 2012. Physico-chemical parameters for testing of water—A review. *International Journal of Environmental Sciences*, *3*, pp. 1194–1207.

Singh, A., Kumar, S. and Nagar, H. 2018. A smart agricultural model by integrating IoT. *International Journal of Pure and Applied Mathematics*, *6*, pp. 729–732.

Srisruthi, S., Swarna, N., Ros, G.M.S. and Elizabeth, E. 2016. Sustainable agriculture using eco-friendly and energy efficient sensor technology. In *2016 IEEE International Conference on Recent Trends in Electronics, Information & Communication Technology (RTEICT)* IEEE Xplore, Bangalore, India (pp. 1442–1446). https://doi.org/10.1109/RTEICT.2016.7808070.

Timsina, J., Wolf, J., Guilpart, N., Van Bussel, L. G. J., Grassini, P., Van Wart, J., . . . and Van Ittersum, M.K. 2018. Can Bangladesh produce enough cereals to meet future demand? *Agricultural Systems*, *163*, pp. 36–44.

Tripathi, A.D., Mishra, R., Maurya, K.K., Singh, R.B. and Wilson, D.W. 2019. Estimates for world population and global food availability for global health. In *The Role of Functional Food Security in Global Health* (pp. 3–24). Academic Press.

United Nations. 2009. Food Production Must Double by 2050 to Meet Demand from World's Growing Population, Innovative Strategies Needed to Combat Hunger, Experts Tell Second Committee. Accessed: Oct. 25, 2022. Available: https://press.un.org/en/2009/gaef3242.doc.htm

United Nations. 2017. World Population Projected to Reach 9.8 Billion in 2050, and 11.2 Billion in 2100. Accessed: Apr. 18, 2019. [Online]. Available: https://www.un.org/development/desa/en/news/population/world-population-prospects-2017.htm

United Nations. 2018. 68% of the World Population Projected to Live in Urban Areas by 2050, Says UN. Accessed: Mar. 15, 2019. [Online]. Available: https://www.un.org/development/desa/en/news/population/2018-revision-of-worldurbanization-prospects.html

United Nations. 2020a. Key Facts-World Urbanization Prospects-Population Division-United Nations. Apr. 2020. Available: https://population.un.org/wup/Publications/Files/WUP2018-KeyFacts.pdf

United Nations. 2020b. 68% of the World Population Projected to Live in Urban Areas by 2050, Says UN|UN DESA|United Nations Department of Economic and Social Affairs. Apr. 2020. Available: https://www.un.org/development/desa/en/news/population/2018-revision-of-world-urbanization-prospects.html.

United Nations. 2020c. Growing at a Slower Pace, World Population Is Expected to Reach 9.7 Billion in 2050 and Could Peak at Nearly 11 Billion Around 2100|UN DESA|United Nations Department of Economic and Social Affairs. Available: https://www.un.org/development/desa/en/news/population/world-population-prospects-2019.html

Villarrubia, G., De Paz, J.F., De La Iglesia, D.H. and Bajo, J. 2017. Combining multi-agent systems and wireless sensor networks for monitoring crop irrigation. *Sensors*, *17*(8), p. 1775.

World Agriculture. 2019. World Agriculture: Towards 2015/2030 by FAO. Accessed: Apr. 15, 2019. [Online]. Available: https://www.fao.org/3/a-y4252e.pdf

World Bank. 2020. World Development Report 2020. Assessed: Dec. 4, 2023. Available: https://www.worldbank.org/en/publication/wdr2020

World Bank. 2022. Climate Smart Agriculture. Overview. Available: https://www.worldbank.org/en/topic/climate-smart-agriculture

World Health Organization. 2018. Global Hunger Continues to Rise, New UN Report Says.

Xu, H., Yu, W., Griffith, D. and Golmie, N. 2018. A Survey on industrial internet of things: A cyber-physical systems perspective. *IEEE Access*, 6, pp. 78238–78259. https://doi.org/10.1109/ACCESS.2018.2884906.

Yeasmin, S., Banik, R., Hossain, S., Hossain, M.N., Mahumud, R., Salma, N. and Hossain, M.M. 2020. Impact of COVID-19 pandemic on the mental health of children in Bangladesh: A cross-sectional study. *Children and Youth Services Review*, *117*, p. 105277.

Chapter 7

Smart aroma culture and effective commercial distillation of major essential oil crops

Md. Fazlul Karim, Soumyabrata Banerjee, Mrinal Kanti Poddar, and Basab Chaudhuri

7.1 INTRODUCTION

7.1.1 Good agricultural practice and some key role players

The world of natural essential oils, oil constituents, their uses in fragrance and flavor industries and innumerable formulation products is fascinating with its ever-expanding potential and opportunities. Several doors are open to enter into this exciting world, and one such challenging gateway at the forefront is aroma culture.

Smart aroma culture or good agricultural practice (GAP) with aromatic plants introduces commercial propagation of a few industrially important aromatic plants as reference crops mainly based on the latest available agro-technologies, duly modified during this work as per requirement, preferably developed, popularized, and established in different agro-climatic zones in India by some premier central institutes and laboratories, dedicated to the field for well over half a century: (a) CSIR (Council of Scientific and Industrial Research)—CIMAP (Central Institute of Medicinal and Aromatic Plants), Lucknow (Head Quarter), UP, and its other centers; (b) CSIR-IIIM (Indian Institute of Integrative Medicine: formerly Regional Research Laboratory), Jammu, J&K; (c) CSIR-IHBT (Institute of Himalayan Bioresource Technology), Palampur, HP; (d) CSIR—IIIM, Jorhat, Assam; (e) CSIR—NBRI (National Botanical Research Institute), Lucknow, UP, and some others.

7.1.2 Secondary metabolites and natural "essential oils"—significance

The so-called aromatic plants, whether natural wild plants or exotic or cultivated botanical species, in their life process produce and store some excretory products (but useful for mankind), called secondary metabolites, in specialized sacs or oil glands of their whole body or body parts like flower, seed, root, bark, leaves, stalk, stem (wood), resin, and epicarp in the form of concentrated oils bearing characteristic aroma and flavor. Such oils are

DOI: 10.1201/9781003495314-7

named after the species or plant and generally known as natural "essential oils" or "aromatic oils." These oils are called "essential" because they are thought to represent the "essence" or odor and flavor specific to each oil having exceptional ability to enter into the human body (say, in the case of aromatherapy) through the skin, olfactory senses, ingestion, or intravenous means and interact with different human body systems to enhance its innate ability and to exert physical and mental well-being or comforts. Thus, aromatherapy, the emerging medium of healing, uses essential oils to tranquilize the body, mind, and soul, leading to calmness, emotional balance, stress relief, and rejuvenation. These are also called volatile or ethereal oils, as they easily evaporate on the action of heat and even at ordinary temperatures. Chemically, these oils are a composite mixture of organic compounds in the group of terpenoids, benzenoides, aldehydes, alcohols, ketones, esters, organic sulfur, and nitrogenous compounds. People belonging to various walks of life and professions starting from medical doctors, naturopaths, aroma-therapists, psychiatrists, cooks to favorites, perfumers use essential oils or oil ingredients by various application techniques to benefit mankind (Dixit, 2019; Karim et al., 2018; Joy et al., 2008).

7.1.3 Uses and historical background

Essential oils are "chemical goldmines" containing about 80 to 400 aroma chemicals. The chemical structures of some of these compounds have been established; most, however, are yet to be elucidated. These oils find use in a variety of industries like cosmetics, perfumery production, pharmaceuticals, food, beverages, and other personal care product manufacturing. Not only making substantial contributions to the national economy, industries using essential oils employ a large number of people like brokers, distributors, processors, and wholesalers in the economic chain.

The use of these oils in health and beauty care is traced back to 1,500 B.C. Fragrances were an integral part of the daily lives of Arabs, who developed distillation techniques several centuries ago. With the advent of civilization, the art and science of perfumery and flavoring have flourished in India since ancient times and perfected inter alia under the patronage of Mughal emperors. In India, Kannauj is considered the Mecca of natural essence, attars, and perfumes, which have a rich traditional history of 1,400–1,600 years. During the last few centuries, the perfumers of Kannauj excelled in the distillation of agarwood, sandalwood, and rose to create the finest scents and attars of the time. During these times, the use of essential oils in therapeutics has become popular in Japan and European countries also (Dixit, 2019; Karim et al., 2018; Joy et al., 2008). Thus, with the passage of time, the industry has gradually diversified into herbs and species, species oils, fragrances and flavors, aromatherapy, neutraceuticals, oleoresins, carrier oils, absolutes, concretes, resinoids, attars, super critical extracts, natural extracts and food colors, and so on (TMV Aromatics Pvt. Ltd., 2023; Ecospice Ingredients Pvt. Ltd., 2023).

7.1.4 World of essential oil crops

Out of 2,50,000 plant species that grow on earth, about 3,000 are of interest so far as essential oils are concerned. Of these 3,000, only about 10% have made mark in world trade. About a dozen major crops contribute to more than 80% in the global market of essential oils.

The most traded essential oils are mints, orange, basil, lemongrass, citronella, geranium, patchouli, eucalyptus, clover leaf, vetiver, sandalwood, agarwood, lavender, jasmine, and tuberose. Other oils in the trade are ginger, cumin seed, dill seed, juniper, nutmeg, rose, turmeric, cinnamon, galangal, sweet flag, black pepper, cardamom, saffron, clove, fenugreek, and celery seed. Some oils and oleoresins are used primarily for flavoring, e.g., ginger, cardamom, pepper, clove, saffron, cumin seed, fenugreek, and celery seed. Oils exported from India to countries such as Russia, the United States, Germany, France, Britain, Australia, the Netherlands, and Gulf are ginger, lemongrass, sandalwood, tuberose, and jasmine. However, some oils like patchouli, lavender, nutmeg, clove, geranium, and rose are still imported from China, Brazil, Turkey, Bulgaria, Australia, Indonesia, and Sri Lanka for meeting the industrial/internal requirements. Moreover, the internal requirements of oils like sandalwood, basil, dill seed, juniper, and mints are fully met from indigenous sources of the country (Joy et al., 2008).

7.1.5 Isolation techniques

These steam-volatile oils are hydrophobic and highly concentrated (e.g., a drop of oil or two are sufficient to exert significant results) and generally extracted by different methods of hydro distillation—e.g., water distillation, water and steam distillation, and live-steam distillation (which are basically governed by twin action of heat and moisture of generated/supplied steam through the charged raw materials involving physicochemical process like hydro diffusion/osmosis, hydrolysis, and decomposition by heat—and by cold expression (e.g., from the epicarp of citrus fruits). Solvent extraction methods are also adopted in some cases of production of "concretes" and "absolutes" of some flowers. Supercritical extraction methods and enflurage techniques are also employed based on needs and demands. The global consumption of these oils and pure oil chemicals or natural isolates is also increasing at ~5% to 15% annually (Dixit, 2019; Karim, 2005; Sanganeria, 2005; Anon, 2014; Guenther, 1949; Lawrence, 1995; Denny, 1990).

7.1.6 Global scenario and india

The bulk of global essential oils are mainly used in industries like spices, processing (for natural isolates/pure aroma chemicals), fragrances (exotic), flavors, and pharmaceuticals. In 2005, the world's total production of essential oils was estimated to be about 1,00,000 to 1,10,000 tons, different "orange oils" constituting ~52%. On the other hand, five major essential oils together occupied ~24% of global shares. These were the oils of the mint family (menthol mint, peppermint, spearmint, bergamot mint, etc.), lemongrass, citronella, patchouli, and vetiver (with India's lone contribution amounting to ~12.64%). The rest of the ~24% shares were held by many other commercially important essential oils. India, then, stood at the number 3 position as a global contributor of about 18% to 19%, and in terms of value, India occupied the number 2 position, particularly for the mint revolution in north India (Sanganeria, 2005).

In 2014, India's production of mint oil (i.e., Mentha arvensis, called menthol mint/Japanese mint/Corn mint) itself was close to 50,000 tons. The production gradually rose from around 6,000 tons in the mid-1990s to ~10,000 tons in 2000, ~12,000 tons in 2005, and ~28,000 tons in 2010, respectively. Thus, in 2014, India already occupied ~80% of the global market as the biggest supplier of oil and oil products (i.e., menthol); here, Uttar Pradesh accounted for ~90% of production and the rest, ~10.0%, were held by Punjab and Rajasthan (Sanganeria, 2005; Anon, 2014).

From the above trends, it is estimated that by the year 2024–2025 India is very likely to produce more or less 1,00,000 tons of various essential oils (out of about 20.0–20.5 million tons of fresh bio-mass processing) bearing about 82–84% shares by above 5-essential-oils. It would involve more or less 7.5 to 8.0 lakh hectares of land for their oil crop (bio-mass) production and another 1.0 to 1.5 lakh hectares of land would also be involved for rest 16% to 18% other essential oils respectively at the grass-root levels. However, most village-level farmers and producer-entrepreneurs at the grassroot levels still face various technical difficulties in terms of huge "process-loss" of essential oils (accounting for ~30% of total internal production per annum, say), batch-to-batch inconsistency in oil yields, oil qualities, production costs, price fluctuation, and pollution hazards due to the inadequacies of different models of "field and steam distillation units" (FDU etc.) in the Indian Essential Oil Industry (IEOI). Here, most of the units are associated with a prolonged distillation periods per individual batch and a huge consumption of fuel, power, man hours, and steam and labour cost. These problems are yet to be overcome/addressed by befitting technology (Karim, 2005; Rajendra et al., 1987; Satchith and Sundaram, 1987; Alkire and Simon, 1990; Varshney, 1992; Tandon, 1998; Kaul et al., 2004).

Synchronization of any "optimal distillation technology/process" with any "suitable device" of distillation is the crying need of the time in India to ensure "future essential oils of purity and consistent quantity" per batch in the backdrop of the following:

(i) Economic sufferings of village-level farmers, producers-entrepreneurs (with most of them unaware of the root causes of sufferings, which lie hidden in the inherent drawbacks of existing FDU systems and associated distillation processes) in the vast IEOI since the 1960s (Karim, 2011; Ozek, 2012) or so.

(ii) Increasing importance and demand for natural essential oils and oil constituents in flavor, fragrance, and processing industries, pharmaceutical industries, aromatherapy, and innumerable fields.

(iii) Global market, which is becoming more quality conscious with the passage of time.

Therefore, for standardization of the "distillation device" and effective processing of a wide range of raw materials for pure essential oils, the above aromatic crops having different ranges of distillation period per batch (e.g., low to high) have been chosen/selected as examples for the present work: menthol mint (Japanese mint/corn mint), lemongrass, citronella, patchouli, and vetiver. Moreover, among these five oils, patchouli oil has been found to have potential antidepressant activity (i.e., neuro-biochemical effects) higher than other oils under this study (Manglani et al., 2011). In addition, it is also extensively used as a fixative in making fragrances.

7.2 OBJECTIVES

The primary aims and objectives of this work were to get rid of all problems faced during 1985–1986 and onward in relation to the distillation of menthol-mint crop in the district of Hooghly, West Bengal, in an entrepreneurship development program, through trial-and-error methods in a breakthrough development.

The aims and objectives of the present investigation are to study the following:

> Comparative efficacy/efficiency of commercial distillation of some five "essential oil crops" in conventional (generally under use in the industry at field levels) and improved devices (synchronized with different optimum production parameters) in the light of (a) fresh feed material/cured feed material and its moisture content, (b) bulk density of charge, (c) fuels under use, their calorific value, and their percent energy utilization, (d) steam rate, (e) distillation speed and rate of distillate, (f) distillation period, (g) energy loss on radiation and other heads, (h) overall yields in comparison to laboratory results, (i) characterization of oils, and (j) setting of optimum parameters for distillation in a novel device for consistent reproducible results (at low cost) in terms of yields and quality.

The essence of this work is, therefore, to prescribe and propose conditions that are industrially conductive for the extraction of essential oils that could be adopted by the IEOI as a whole.

7.3 RESEARCH METHODOLOGY

7.3.1 Raw materials used

In order to produce *five types* of raw materials (discussed above) at different sites from nursery to semi-commercial levels of *elite* planting materials (i.e. 14-cultivation varieties, corresponding to "5-aromatic crops") were collected from time to time from authentic sources. Nurseries and small plantations were raised, and crops were grown thereafter (by standard agro technology mainly described/circulated by CSIR-CIMAP, Lucknow, in their different farm bulletins, other publications, and acquired traditional knowledge) at three different sites in *West Bengal* in the plains: e.g., at Bandpur, Hooghly; at Deganga, 24-Parganas (North); and in the hill areas of Mungpoo, Darjeeling, respectively.

Cultivation involved the following basic steps:

(a) Preparation of land and layout
(b) Application of manures and fertilizer (as basal dose and top dressing)
(c) Sowing of suckers/seedlings, rooted cuttings, slips, etc.
(d) Gap filling
(e) Weeding, hoeing, and interculture
(f) Disease and pest control, if any
(g) Harvest of biomass/digging of roots
(h) Post harvest management

In the hilly areas of Mungpoo, citronella, lemongrass, and vetiver were grown in fallow lands/virgin soil under wild conditions mainly (i.e., without use of any chemical fertilizer, but other steps of propagation remaining the same). No threats of disease and pests were found there for the abovementioned crops.

Common things in the cultivation are sandy loam soils, rich in humus, having pH 5.5 and above; these are suitable for the crops under discussion. The hilly areas of Darjeeling, up to an altitude of ~1200 m, are found suitable for all the five crops. In the plains, the area of plantation should not be prone to waterlogging and should have irrigation facilities. In Darjeeling, plantations are raised based on yearly, average well-spread rainfall, ranging from 272 cm to 384 cm, throughout the year. All the crops under discussion respond very well to farm yard manure (FYM) and the micro nutrients available in the market. The general rule is that crops like citronella and lemongrass should be harvested 2″–3″ above the "knot" or branching point of the slips (to prevent immature death of the slip). Heaping of all fresh "BIOMASS" after the harvest must be avoided (to prevent heat-generation within the "heap" and remarkable loss of oil due to fermentation), and all raw-materials, e.g. menthol mint, citronella, and lemongrass, should be distilled, preferably on the same day of harvest (as instant distillation would save more or less 10% of oil from getting lost) or if not possible on the day, crops should be well spread over under shed and distillation should be completed by the following day. Lastly, none of the five crops are grazed by cattle normally.

7.3.1.1 Menthol mint (mentha arvensis l.; Family lamiaceae (labiatae)) cultivation

Five species of the mentha family are commercially cultivated throughout the world since a very long period; out of these, four species are immensely important in the essential oil industry. The four species, respectively, are (1) Japanese mint or menthol mint or corn mint (*Mentha arvensis*), containing 70 to over 80% menthol; (2) peppermint (*Mentha piperita*), flavored for total oil sweetness; (3) bergamot mint (*Mentha citrata*), rich in linalool; and (4) spearmint (*Mentha spicata*), enriched in carvone (60% or so). The former two species of mentha were introduced in India by the Regional Research Laboratory (RRL), Jammu, CSIR (now CSIR-IIIM, Jammu), in 1954 with a view to propagate their cultivation, leading to the mentha-based essential oil industry in India. Sometime later, the Central Indian Medicinal Plants Organization (CIMPO), a National Laboratory under CSIR (established in 1959), now known as CSIR-CIMAP, came forward with the same task to popularize such crops among the rural farmers of the Tarai region of U.P., Jammu and Kashmir, etc. and to develop high-yielding varieties. So, the history of mentha is ~7 decades old in India. And nowadays farmers or growers of UP (Tarai region), Punjab, Haryana, Jammu and Kashmir, Himachal Pradesh, Madhya Pradesh, Gujarat, and Bihar have commercially taken up such crops for cultivation as cash crops and thus fetch good profits. Therefore, the mentha-based essential oil industry is now well established in such areas, and the cultivation of mentha species has extended in India to about one lakh hectare of land in 1998, which led to the production of about 12,000 MT of mentha-based oil or mint oils. India stood first, for the first time, in the production of menthol-mint oil, beating neighboring China. However, now in 2022–2023, only Menthol-mint cultivation is extended to about 4.0 lakh hectares (ha) of land to produce about 45,000 MT of mint oil, giving rise to ~34,000 MT of menthol for its internal consumption and exports.

In the present work, vis-a-vis in an "entrepreneurship development programme" entitled "Cultivation of mint crop for production of mint oil and processing of mint oil for production of menthol," the first project report dated 12 June, 1984, was prepared and posted to the entrepreneurs by late *Dr. S. C. Datta*, founder director of CIMPO (now CSIR-CIMAP), Lucknow. Then, after registering the SSI (Small Scale Industry)-Project with DIC (District Industries Centre) at Hooghly, suckers (i.e., roots) of MAS-1 variety of Menthol-mint, a seasonal crop of 6–7 months duration, was collected in

February 1985 from the Pantnagar Unit of CIMAP (CSIR), with the core guidance of late Dr. Datta. Thereafter, a nursery was raised at Bandpur, Hooghly, West Bengal, on a sandy loam soil having a pH value of 6.2 and contents of total nitrogen, total phosphorus and organic carbon as 0.11%, 0.27% and 0.28% respectively. After applying the prescribed basal dose of FYM and NPK-fertilizer, urea was applied as top dressing from time to time with adequate care and culture. The nursery was thus carefully maintained and afterwards multiplied with cuttings in different other plots during August-September, 1985. Propagation of the crop was finally carried out with suckers during January 1986 by prescribed method of CSIR-CIMAP (Farm Bulletin, 2004). Luxurious harvests started to come up during the months of April–May 1986 and batches of distillation were conducted in a direct fired "Field Distillation Unit (FDU)" (as was commissioned during the second week of March, 1986) one after another. Side by side, distillation of representative samples of Menthol-mint crop was carried out in the laboratory (Pharmaco-Anatomy Laboratory) under Late *Professor Prafulla Chandra Datta*, Department of Botany, University of Calcutta to ascertain actual oil percentage [as was found 1.0% (v/W), on an average] of the crop.

But due to inherent drawbacks/limitations of the said FDU-System the programme suffered from serious setbacks in terms of yield, oil quality, time consumption, production cost and environmental pollution etc. Initial *process-loss* went upto ~70%.

However, all the above-mentioned problems could be resolved by tedious "trial and error efforts" during next 3–4 years period with a breakthrough development resulting in an "optimized system of essential oil distillation," termed here as Modern Distillation Plant (MDP-System: where an Optimized Process of distillation was coupled with the said Modern Device). It was granted by an Indian Patent in 1998 (Patent No. 199380). Needless to mention that past work is now the foundation of this chapter.

Later, other elite varieties of Menthol-mint suckers, such as Shivalik-88, Himalaya, and Kosi were successfully taken into propagation practices mainly based on the manual (farm-bulletin) of CSIR-CIMAP and traditionally acquired experience. Nowadays, CSIR-CIMAP has released a new cultivar: CIM Unnat, which has also been taken into cultivation practice under this work.

The crop is found to grow well in different agro climatic conditions of West Bengal, both in the plains and hill areas of Darjeeling upto an altitude of ~1200 M from mean sea level.

7.3.1.1.1 Salient features of the crop and results

(a) *Propagation*: Propagation procedure resembles "potato cultivation" where ~50–100 kgs of healthy suckers are required/acre by seedlings (30–40 days old) and direct sowing methods respectively; spacing of seedlings is 15 × 45 cm. (b) *Recommended dose of FYM* and *NPK fertilizer*: FYM at

6.0 MT/acre is applied during land preparation along with total NPK at 150:60:40 for 2-harvests [1/3rd N and total P and K are applied as basal dose and rest N in splits as top dressing]; for a single harvest total NPK dose is 80:40:40. (c) *Biomass* and *Oil*: 1st harvest is done at flowering stage of the crop; 6–10 MT of fresh biomass is available/acre in1–2 harvests during 6–7 months period with 50–70 kgs of oil yields. (d) *Oil percentage*: Oil% is found to be 0.4% to 1.0% (depending on the cultivar). (e) *Average distillation period/batch*: 1.0 hour (in MDP-System) to ~4.5 hours—6.0 hours (in most of the conventional FDU-System). (f) *Uses*: Isolation of Menthol and De-mentholized Oil from the bulk-oil; useful in industries like pharmaceutical, paan-masala, tobacco, toiletries, bakery/confectionery, beverages, cigarette, fragrance and flavor, etc. (g) *Principal constituents*: Menthol, Menthone, Menthyl acetate, Neo Menthol etc. (h) *Principal producing countries*: India (covers ~80% of Global Production), China (~10%), Japan (~10%). (i) *Producing States in India*: U.P. and Bihar (producing ~90%), rest by Punjab, Haryana etc. (j) *Generated paid man-days/acre*: ~80–125 days. (k) *Notes*: (a) More profitable *after harvest of potato* in West Bengal (vast scope is there after potato, but it is still almost untouched); (b) Market price of the oil highly fluctuates from ~Rs.1000–1200/-per kg in Kolkata (Rs.1100–1300/- per Kg [Shivalik]: UnivDatos Market Insights Pvt. Ltd., 2022; Dubey, 2022).

7.3.1.1.2 Mint-waste as an alternative energy source and revenue

Calorific value of dried waste of Shivalik-mint is found to be 4093.79 Kcal/Kg (~64.34% of Grade 2-B steam coal). There is extensive possibility to operate *Briquetting Plants* around the distillation-units in the *Mint-Belts* (the hub of distillation-wastes) of the country. If the so-called MDP-System of distillation can be put into operation there, using briquettes as useful fuel, over 50% of briquettes will be surplus after each batch of distillation. Such surplus can be a huge source of revenues if sold in the market and cost of distillation and, even, *modernization* will be remarkably balanced from the revenue so earned.

Example: Vast Menthol-mint industry itself produces over 6.0 million MT of fresh biomass for distillation as on date. Therefore, distillation wastes of Mints there can give rise to over 1.1 million MT of good quality *briquettes* and after distillation with such briquettes, there will still be a surplus of about 0.55–0.6 million MT of briquettes and that to earn additional revenue.

7.3.1.2 Lemongrass (cymbopogon flexuosus) cultivation

Lemongrass (cymbopogon flexuosus, family Poaceae) is a commercially important aromatic grass, semi-perennial in nature and its cultivation was initiated in Kerala over 100 years ago as “Coachin grass.” Nowadays, it is cultivable in different agro-climatic zones of the country with commercial success. The oil of Lemongrass (also called Coachin oil) has a great demand in the country itself and in abroad. It is rich in citral contents (75–87%) which in turn is the precursor for production of "alpha and beta" ionones and, thereby, for the synthesis of Vitamin-A (from beta-ionone). The oil as such bears bactericidal and insect repellent properties also.

Because of poor performance of "Coachin grass" in terms of commercial return and oil yield per acre, various high yielding strains of the crop e.g., RRL-16, CKP-25, OD-19, Pragati, Praman, Cauveri, Krishna etc. have been developed and introduced to the farming community from time to time by different National Laboratories (Singh and Singh, 1998; Bhan et al., 2001; Kaul et al., 2004). A new cultivar, CIM Sekhar, has now been introduced by CSIR-CIMAP, whereas other 2-cultivars L-8, L-14 have been recently introduced by CSIR-IIIM, Jorhat in North Eastern region.

Cultivation and/or distillation of the crop have been discussed by various workers in different publications from time to time (Singh and Singh, 1998; Bhan et al., 2001; Kaul et al., 2004; Satchith & Sundaram, 1987; Karim, 2000, 2005).

Economic potential of the crop has been studied very well by the author over a period of 25-years in plains and hill areas (in virgin land without application of FYM and chemical fertilizers) of West Bengal. Cultivation of Lemongrass under irrigated conditions with application of FYM at 12MT/acre; NPK-fertilizers at 60–110 kgs N, 16–32 Kgs P2O5 and 16–40 Kg K2O and Zinc sulphate at 8 Kg/acre respectively has been reported (Singh and Singh, 1998; Bhan et al., 2001).

World production of Lemongrass oil is estimated to be 300–350 MT/year against 200 MT annually by India (Joy et al., 2008). As of now, global production is reported to be ~1500 MT against demand of 5000 MT and India's production of 700 MT per annum respectively.

7.3.1.2.1 Briquetting from lemongrass waste

That Lemongrass is a potential alternative fuel in the form of "briquettes" having 9.89% moisture, high Bulk-Density (1290.0 Kg/M^3) and calorific value of ~3466.6 Kcal/Kg. and how such "briquetting" can be done on commercial scale at low cost have been described by other workers (Paul, 2013). However, the author has found out the calorific value of distillation waste of Krishna variety of Lemongrass which is 4296.15 Kcal/Kg (i.e., ~67.5% of Steam Coal: Grade 2-B).

7.3.1.2.2 Salient features of the crop and results

(a) *Commercial life*: Semi-perennial grass having commercial life: 5–6 years (in the plains) and 12–20 years in hill areas of Darjeeling. (b) *Planting materials*: 12,000–20,000 slips for ~11,000 healthy clumps per acre (including gap filling) with spacing of 45 cm × 60 cm in general; [however, ~25,000 clumps/acre has also been reported (Bhan et al., 2001). (c) *Quantity of biomass/*

acre: ~15 to 18MT per acre per year, an average of 5-years. (d) *Yield of essential oil/acre*: ~80–100 kgs oil, bearing an oil % of 0.6 to 1.0 % (based on the cultivar and season). (e) *Distillation period*: Average 75 minutes/Batch (MDP-System) and 4–6 hours/Batch (in conventional FDU-System). (f) *Uses of the oil*: For producing aroma chemicals, such as Citral-a, Citral-b, Geranyl acetate etc. by "fractional-distillation method" and, then, perfumery chemicals—alpha ionone, beta ionone, synthetic Vitamin-A etc. Also used as such in perfumery, cosmetics, soap industries and aromatherapy. (g) *Principal oil constituents*: Citral-a, Citral-b, Geranyl acetate, Linalool etc. (h) *Principal producing countries*: India, Sri Lanka, Burma, Thailand etc. (i) *Principal producing states in India*: Kerala, Karnataka, Tamil Nadu, Uttar Pradesh, Uttarakhand, Assam, Arunachal Pradesh, West Bengal etc. (j) *Yearly production in India*: ~700 MT (reported). (k) *Generated paid man-days/acre*: ~150–180 numbers per year. (l) *Other features*: (a) Vast scope of the industry is there in West Bengal; (b) Ample of process-loss of oil is there due to post-harvest mishandling of biomass and limitations of FDU-System in general; (c) Market price fluctuates very much from ~Rs.1200–1600/- per kg of oil in Kolkata (Rs.1800–2200/- per kg; Dubey, 2022).

7.3.1.3 Citronella (cymbopogon winterianus) cultivation

Citronella Java Type grass is botanically known as "Cymbopogon winterianus Jowitt, family Poacae." It was introduced in India by Regional Research Laboratory, Jorhat (now, CSIR-IIIM). It is an ideal crop for cultivation in Hilly slopes. The advantages of raising this grass in hilly slopes are:

1. It does not require terracing (raised level place).
2. It can be grown under rain-fed conditions in Darjeeling hill as in N.E. India.
3. Once planted, it yields grasses for 5 to 6 years without replanting in the plains of West Bengal; whereas, it is found to give commercial yields of biomass for over 20-years in hill areas of Darjeeling.

In the mid-1990s, India's consumption of citronella oil was above 650 MT per annum. Taking into account the normal increase in consumption of products derived from citronella oil, the consumption was estimated to be above 1,500 MT by the year 2000. However, during the year 2004–2005, the northeastern region accounted for the production of 150 MT of the oils per annum (Jallan, 2004). But, the production of Java oil in India was, later, reported to be 300–350 MT/year (Joy et al., 2008).

7.3.1.3.1 Planting—material

Only authentic and improved varieties of planting—materials, viz. Jorlab C-2 and Bio-13 etc., should be used for cultivation. Ceylon Citronella (also called Bangla variety) which is found to be hardy has also been taken into cultivation practices by the author. These varieties yield increased output of grass and more oil. These strains are found to be more resistant against attacks of insect, pest, fungus and other diseases resulting into increased longevity of the plantation and higher returns from the crop.

The grass propagates well through splitted slips, the planting—materials, in a variety of climatic and soil conditions. The slips should be planted preferably during the onset of monsoon and even in spring. Spacing generally followed: 45–60 cm × 60 cm in between slips and rows. Therefore, one-acre of land (4047 sq. meters) would well provide ~11,000 planting holes, generally 15 cm deep, for planting of slips and in each of the hole two healthy slips should the placed. It takes almost 30 days to establish on the soil, thereafter tillering (branching) starts. However, survival rate of so planted slips generally found to be more or less 75% while planted in dry season. Gap filling should be done at 30–40 days after transplantation. However, success depends on the care it receives in the first year in respect of weed control, manuring, inter-culture, proper irrigation etc. and it must be protected from water logging of the plots; or it may be planted on ridges to protect it from water logging.

The first harvest can be taken 3–4 months after planting the slips and thereafter every 45–75 days depending upon the soil fertility, climatic factors and management practices. Under favorable conditions at least 5–7 harvests are possible. It should be borne in mind that the harvest must be done always above the "knot," the branching point of the stem. Furthermore, harvests of too young leaves appear to reduce the geraniol content of the oil, whereas harvest beyond 90-days affect the total yield of leaves due to partial drying, weedy stems etc.

7.3.1.3.2 Fertilizers

In hilly and virgin soils, it is not necessary to add any fertilizers during the first few years of cultivation. Later on, if required the usual granulated NPK mixture can be used. It should be applied one week after every harvest. The total quantity required per acre per year is about 200 to 250 kgs.

7.3.1.3.3 Weeding

During first six months after planting proper care should be taken to remove weeds, subsequently weeding/hoeing may be undertaken after harvesting, say 2 to 3 time in a year.

7.3.1.3.4 Briquettes from distillation waste of citronalla

There is extensive briquettes making possibility with the distillation waste of Citronella also, having calorific value of 4089.04–4266.14 Kcal/Kg (accounts to ~66.0% of Steam Coal: Grade 2-B) of dried waste, which can be used in the boiler for distillation of the grass for cost reduction on fuel head. Surplus briquettes, once sold, would balance the cost of man-power on distillation head and Optimization cost, if any, of the conventional FDU.

7.3.1.3.5 Salient features of the crop and results

(a) *Quantity of Biomass/acre*: ~15–25 MT of fresh Biomass per acre per year, an average of 1st 5-years depending on the cultivar. (b) *Distillation period/Batch*: Average 2.0 hours/Batch (in MDP-System); 3.5–8.0 hours in most conventional FDU having some process-loss. (c) *Oil yield/acre*: ~90–105 kgs oil/acre per year with oil (%) of ~0.45%—0.65% (for Bangla variety) and ~0.7%—1.3% (for Jorlab C-2 variety) in general in different seasons. (d) *Uses of the oil*: Extensively used in preparation of room freshner, phenyl, mosquito repellent, soap, detergent, cosmetics, food and beverages, and in perfumes. Also used for isolation of aroma chemicals. (e) *Principal oil constituents*: Citronellal (3–6%), Geraniol (15–23%), Citronellol (3–9%), Limonene (7–12%), Methyl isoeugenol (7–11%) [In *Ceylon variety*]. Citronellal (31–39%), Geraniol (20–25%), Citronellol (9–13%), Geranyl acetate (3-6%), Limonene (2–5%) etc.[in Assam variety, i.e. *Jorlab C-2*]. (f) *Principal producing countries*: Sri Lanka, Indonesia (Java), China, Taiwan, Guatemala and India. (g) *Principal producing States in India*: Andhra Pradesh, Kerala, Karnataka, Tamil Nadu, Orissa etc. (h) *Yearly production in India*: 300–350 MT in 2008 and at present (?). (i) *Generated paid man-days/acre/year*: ~160–210 numbers. (j) *Notes*: (a) Ample of scope of farming by self-help Groups in West Bengal; (b) Ample of mishandling of biomass prior to distillation and limitations of most FDU-Units render remarkable process-loss of valuable oil; (c) Market price fluctuates too much from ~Rs.1100–1600/- per kg of oil in Kolkata (Rs.1600–2000/- per kg.; Dubey, 2022).

7.3.1.4 Patchouli (pogostemon cablin) cultivation

The patchouli plant [Pogostemon cablin (Blanco) Benth; also Potchouli] is an aromatic bushy and well branched shrub (herb) belonging to the genus Pogostemon and mint family Labiateae with fragrant leaves bearing medicinal and aromatic properties. It grows to about two feet to three feet bearing small pale pink-white flowers and is semi-perennial in nature under cultivated conditions (Akhila and Tewari, 1984; Ramya et al., 2013). It is a shed loving plant (~40% shed is ideal for growth of the crop) and a native to Philippine and broadly found in the subtropical Himalayas, South East Asia and far East (China, India, Taiwan, South Korea, Singapore, Australia, Japan). Because of synthesis of a composite mixture of fragrant secondary metabolites in its leave, in particular, called essential oil of patchouli (Yahya and Yunus, 2013), commercial cultivation of the plant has been taken up in many countries around the globe, viz. Indonesia (Java, Sumatra), Malaysia, China,

Singapore, Hanoi (Vietnam), West Indies, Mauritius, Thailand including Brazil and India (Akhila and Tewari, 1984; Ramya et al., 2013; Lawrence, 1981; Bunrathep et al., 2006). The thick and light yellow to dark brown oil is recovered from the Shed-dried and well cured whole plants/leaves of patchouli by hydro-distillation/steam distillation methods (Ramya et al., 2013; Bunrathep et al., 2006) and the oil bears well acclaimed characteristic musky-sweet/strong spicy notes and long lasting herbaceous, woody, earthy, camphoraceous smell (Ramya et al., 2013).

Once planted, patchouli can be maintained commercially for three years in general. Acidic soil having pH value of 5.5 to 6.2 is described ideal for the crop. On maturity it attains a height of about 0.5 M to 1.2 M (Ahmed, 2005). Commercially important cultivars, now cultivated in India, are: Johore, Indonesian and Malaysian types, CIM-Shrestha (by CSIR-CIMAP) and Kelkar type (Keva-Kelkar: selected from the natural flora by Kelkar Scientific Research Centre, which is not a genetically modified material) (Ahmed, 2002; S.H. Kelkar & Co. Ltd., 2002).

The oil is rich in patchouli alcohol and it blends well with many essential oils and is almost a perfume itself on aging (Varshney, 1992; Singh, 1998). On yield percentage, quality, and odor point of views, oils from the first three cultivars rank as Johore>Indonesian>Malaysian (Ahmed, 2002). The oil is rich in woody, earthy, and fruity odor (in top notes) as well as herbaceous odor (in dry out note) (Maheshwari et al., 1993).

The land is generally tractorized (ploughed) twice at an interval of 5–10 days. Before final ploughing and levelling, *FYM* (at about 6.0 MT/acre), *anti-nematode materials* (e.g., organic pesticide "nematex" at 12–15 kg/acre or Furadan at 10.0 kg/acre, where the former is preferred to Furadan for better persistence and results) and basal *NPK fertilizer* (at 0:20:12 per acre) along with *micronutrients* for soil application (at 6.0 kg/acre) are applied uniformly on the surface. Irrigation channels are then laid out and the land is kept as such for two days in general. Previously prepared ~30–35 days old rooted cuttings of patchouli are then transplanted at a spacing of 45 × 60 cm (in the afternoon session in case of sunny days or at any time on cloudy days), followed by light irrigation, and it results in about 13,000 population/acre. The moisture level in the soil is carefully maintained thereafter. Light to moderate earthing is done in rows of patchouli during 21 to 35 days after the first weeding is completed, and this is followed by irrigation. The plantation is kept weed free by 2–3 weeding and hoeing during the establishment period and 1–2 light weeding during the rest of the year. Total *nitrogen* (at 60.0 kgs/acre) as urea is then applied (duly mixed with 2.5 to 3.0 times soil or composed two days prior to application) in 5–6 equal splits on, say, 45 days, 90–100 days, 135–145 days after transplantation and 40–50 days after the first, second, and third harvests during the year. A top dressing of *muriate of potash*, K2O (i.e. K at 8.0 kg/acre), is done during 90–100 days along with urea to provide the crop an additional resistance against adversity. As *foliar spray*, a solution of 0.5% urea in combination with 0.25%

micronutrients is applied 20 days and 10 days prior to each harvest. Foliar spray of *fungicide* (0.1% bavistin) is done after each harvest as a precautionary measure, and 0.25%–0.30% of fungicide is also sprayed when there are symptoms of fungal infestation. Uprooting of the affected plants is done immediately. Distillation of well-matured patchouli has been done through an advanced distillation technology with a yield of 1.6%–4.7% (w/W) containing 27–39% patchouli alcohol (Karim, 2006).

7.3.1.4.1 Salient features of the crop and results

(a) *Crop type and life span*: Semi perennial; commercial life of 1–3 years. (b) *Planting materials*: Rooted cuttings (~30–35 days old); ~15,000 rooted cuttings per acre for ~13,000 healthy bush (to be planted, except during hot summer and winter season, at spacing of 45 × 60 cm). (c) *High-yielding cultivars*: Keva-Kelkar, Johore (Indonesia), Indonesian, Malaysian, CIM-Shrestha (CSIR-CIMAP). (d) *FYM* and *NPK fertilizer, care*, and *culture*: As prescribed/as per requirement. (e) *Quantity of harvested biomass/acre*: From about 14 MT, 15 MT, and 13 MT of *fresh biomass*/acre during the first, second, and third year to about 2.5 MT (first year), 2.75 MT (second year), and 2.3 MT (third year) of well-cured and matured (for 3.5–4.5 months) *whole plants*/acre per year. (f) *Distillation period*: 3.5 hours average/batch (in MDP-System); 8–24 hours (in most conventional FDU systems). (g) *Yield of essential oil/acre*: ~38–50 kgs/acre in general (with 1.6–4.7% yields in optimized unit based on the oil contents in the cured mass). (h) *Uses*: In aromatherapy, mosquito repellent, food, drinks, perfumes, cosmetics, soap, zarda, pan-masala etc.; also used as fixative. (i) *Type of essence and note*: hard and sweet, characteristic note of patchouli; sweet note of oil increases with aging. (j) *Principal oil constituents*: patchouli alcohol, alpha guaiene, alpha patchoulene, beta patchoulene, alpha bulnesin, seychellene, norpatchoulenol, eugenol, pogostone, pogostol etc.; content of patchouli alcohol in the oil (~26% to 33%) is a measure of pricing of the oil. (k) *Principal producing countries*: Indonesia, Malaysia, Singapore, China, South Africa, Vietnam, and India. (l) *Principal producing states in India*: Tamil Nadu, Karnataka, Assam. (m) *Yearly estimated production in India*: ~100 MT. (n) *Yearly estimated consumption in India*: ~700 MT (i.e., more or less ~600MT of oil is imported to India). (o) *Generated paid man-days/acre per year*: ~275–335 numbers. (p) *Notes*: (a) Well-dried and tightly packed cured leave is stored under shed for 3.5 to 4.5 months prior to distillation to obtain well-matured crops and oil; (b) high scope of farming is there in West Bengal also (through small farmers, self-help groups etc.); (c) imported quality oil can be produced in the country under optimal conditions using optimized MDP-Unit; (d) India has the potential to meet its own internal demands by its own production with proper planning; (e) market price fluctuates very much from ~Rs.4,000 to 4,500/- per kg of oil in Kolkata (Rs.5,800–6,200/- per kg; Dubey, 2022).

7.3.1.5 Vetiver (vetiveria zizanioides) cultivation

Vetiver (Vetiveria zizanioides (L.) Nash) is a semi-perennial grass belonging to the family "Poaceae." It is an important aromatic plant. Its genetic name is derived from the Tamil word "Vetiveru." The roots of the plant have been employed to make baskets, curtains, and mats and in the production of essential oils and medicines since time immemorial. The oil of vetiver is widely used worldwide in high-grade perfumes, cosmetics, chewing gum, tobacco, soft drinks, fragrance, flavors, attar, soaps, and many industries. In India, vetiver is mainly grown in the states of Uttar Pradesh, Rajasthan, and the southern states. The roots of both wild and cultivated plants are used for extraction of the oil to partially meet the demand of the country. It can be cultivated in a wide range of agro-climatic conditions as irrigated and non-irrigated crops under tropical and subtropical weather.

Vetiver is cultivated by splitted slips (about 15,000 slips are required/acre plantation having spacing of 30 × 45 cm in general) preferably during January–February and July–October. The process of cultivation, care, and culture all resemble lemongrass and citronella cultivation. It is a semi-perennial (*~10–30 months*) crop. CSIR-CIMAP has released various high-yielding cultivars, namely Imran et al. (2022), Dharini, Gulabi, Kesari, CIM-Viridhi, and CIM-Khusnalika. CIM-Viridhi matures in 10–12 month period. The application of NPK-fertilizer is necessary for optimum growth of the plant and root yield. The recommended dose of NPK is 50:24:20 per acre of plantation. One-fifth dose of N and total P and K are applied as basal dose prior to transplanting; rest of N is applied in four splits after 30 days, 60 days, 90 days, and 120 days of establishment of the crop by broadcasting under optimum moisture. Gap filling is done after 30–40 days, with subsequent replacement of dead plants by the established new slips. Digging of root after 10–12 months is recommended for winter-planted crops. December/January is found to be the right time for harvesting operation, as the oil content is maximum during that period. CSIR-CIMAP has developed an uprooting machine "CIM-KHUS-Upay" that reportedly reduces the uprooting cost by ~20%.

The process of distillation consists of proper chopping of the clean roots to 5–10 cm long followed by soaking with water prior to charging into the still. This practice would help accommodate more roots and accelerate the process of hydro-diffusion, osmosis, and other phenomenon during distillation. Distillation demands high skill and experience on the part of the "operator" of the UNIT for optimum recovery of the oil having its specific gravity almost equal to that of water. Oil recovery on the commercial scale ranges from 1.2% to ~2.0% (Farm Bulletin, 2007).

World production of vetiver oil is approximately 250–300 MT/year, where India contributes only 20–25 MT (Joy et al., 2008; Farm Bulletin, 2007). India's production is far below its indigenous requirement. Oil of north Indian origin is considered best in the global market.

7.3.1.5.1 Scope of briquetting

There is a high scope of briquetting with the available aerial biomass (green herbage) of the crop; apart from its "root," which is used for essential oil production, it also provides green herbage at regular intervals during the year. The available green herbage can be used as cattle feed and commercially utilized after proper drying for briquetting purpose. Dried herbage has a calorific value of 4,068.52 kcal/kg (~64.0% of steam coal: Grade 2-B). The briquettes so produced could be the effective fuel for distillation of the root to balance the overhead cost and/or to earn additional revenue. Therefore, dried herbage is not a waste provided one takes care of the *heat* of this "waste."

7.3.1.5.2 Salient features of the crop and results

(a) *Quantity of harvest*: About 8–12 quintals cleaned and cured roots/acre. (b) *Distillation period*: 5–7 hours per batch (in MDP-System); 12–32 hours/batch in most conventional FDU-units with huge process-loss. (c) *Yields of essential oil/acre*: About 8–15 kgs. (d) *Uses of the oil*: In perfumes, cosmetics, soft drinks, tobacco and zarda and in many more industries (as fixative and natural essence). (e) *Type of essence/Note*: Soothing and lasting scent with khus note, rose note and saffron note etc. depending on the cultivar. (f) *Principal constituents*: Vetiverol, alpha-vetivone, beta-vetivone, vetiveryl vetivenate, Khusol, Khusimol, Khusitone etc. (g) *Principal Producing Countries*: Indonesia, Italy, Reunion island, Haiti, Thailand and India. (h) *Principal Producing States in India*: Karnataka, Tamil Nadu, Kerala, Orissa, Madhya Pradesh, Uttar Pradesh, Uttarakhand and Rajasthan etc. (i) *Generated paid mam-days/acre*: 225–300 numbers approximately. (j) *Notes*: (a) The crop protects soil erosion; (b) India has the potential to meet its own internal demands by its own production with proper planning; (c) Market price per kg of oil Rs.45,000–48,000/- (for Ruh Khus vetiver oil) and Rs.26,000–30,000/- (for South Indian vetiver oil) respectively (Dubey, 2022). This way all available raw materials from five foundation crops were produced and subjected to distillation in different forms—e.g., fresh, wet, semi-dry, dry, and cured form.

7.3.2 Fuels used

Sources of energy for generating required steam (which can be: wet, dry, semi super-heated steam etc.) were: Grade-2B Steam Coal, fire-wood, electricity, Kerosene and dried distillation wastes (called "spent herbs") etc. having different calorific values.

7.3.3 Distillation equipment and plants

Distillation of different raw materials in various batch sizes (i.e., laboratory-scales to commercial-scales) was carried out by 3-methods of hydro-distillation at 3-sites of West Bengal in 9-sets of "Equipment and Plants" (including a Clevenger apparatus) consisting of:

(a) 3-Non IBR vertical type Steam Boilers and 1-Water-boiling vessel;
(b) 12-Mild Steel (M.S.)/Stainless Steel (S.S.) Distillation-Still(s) and 1-"Round Bottom Corning-Flask,"
(c) 11-Condensers and (d) 9-Water-oil separators (Receivers).
- *Additional components*: (i) Heating mantel, Heating Element and Kerosene Stove (for Laboratory and bench-top equipment), (ii) Water Reservoir, (iii) Cooling Tower and Water Pit, and (iv) Tools and tackles.

Eventually, an optimized *Modern Distillation Plant* (MDP-System) was emerged with following features and components to handle different aromatic crops on commercial scale:

7.3.3.1 Features

1) A single centrally located MDP-Unit is compatible to multiple types of aromatic crops.
2) A single MDP-Unit more effectively and economically takes up the work-loads of 3–5 "Conventional Distillation Units"/shift, known as FDU, in prevailing industry (with ~Zero Process-Loss).
3) Batch size: Upto about 2,000.00 Kgs of fresh Mints.

7.3.3.2 Basic components

(1) One steam-generator.
(2) 2-Vessels or Stills (for alternate and continuous batch operations, even for round the clock operation).
(3) 1- Condenser (Common).
(4) 1- Water-Oil Separator (coupled with Receiver).

7.3.3.3 Additional requirements

(a) 1-Cooling Tower (Optional)
(b) Overhead "water reservoir"
(c) Water holding PIT
(d) Briquettes making Machine (Optional)

7.3.4 Distillation of essential oil crops

In the present set of experiments major shortcomings of a conventional "direct fired type Water and Steam Distillation plant" [popularly known in the industry as the "Field-Distillation Unit (FDU)"] were eliminated and substituted by effective "hydro-distillation" (i.e., water distillation, water and steam distillation and live-steam distillation) using 5-different aromatic raw-materials under the trials. These experiments were repeatedly carried out starting from "laboratory scale"/"pilot scale" to "commercial scale" in 9-different duly optimized "equipment sets and plants," duly coupled with "suitable process," named MDP (Modern Distillation Plant)-System, based on the principles of "hydro-distillation" (Guenther, 1949; Lawrence, 1995; Denny, 1990). Here, each crop-material was evenly and tightly packed (to avoid steam channeling) in duly *insulated* distillation stills and, then, covered with the lid. Packed crop-material was then digested under "optimal conditions" (Karim, 2011) by desired quantity of generated steam preferably at *~0-psig* (i.e., at ~atmospheric pressure) to *~12 psig* to start distillation. The generated "mixed vapor" was then passed through a "condenser," and extracted "oil" was automatically collected in the receiver along with the outgoing condenser "distillate" at "optimal rate" (as was set in "ml" per liter of load-volume in the still per hour). Thus, complete distillation to took place at the desired and lowest contact times of steam with the raw material (where the "distillate-rate" was controlled by steady and proportionate "steam-supply rate"). This led to superior quality yields. Finally, the resulting oils were characterized by GC/MS method.

7.3.4.1 Phenomenon (involved)

- Distillation of essential oil crops is basically governed by *the twin action* of heat and moisture of supplied/generated steam
- Optimal distillation process, conducted at a low cost, was found to be a function ("f") of a, b, c, d, e, f, g. (represented as f (a,b,c,d,e,f,g . . .)), where

  a = bulk density (BD) of loaded crop and prevailing physical state of the raw material;

  b = steam pressure of boiler;

  c = steam quality (wet, dry, semi-super-heated, etc.);

  d = distillation temperature and pressure within the vessel (still) at "steady-state" of ~*96*°C (at~1200M altitude/0-psig)—*118* °C (~12-psig)/*102* °C (in most cases);

  e = crop-specific steam dose/boiler HP;

  f = minimal radiation loss;

  g = optimal system design for optimum "distillation speed" or adequate material-energy balance in terms of distillate (e.g., at 90±30 ml/hour/liter of load-volume of the still (depending on the type/physical state of raw-materials and batch-sizes or still-volume)), etc.
- Each component of an optimized device, irrespective of batch sizes, is well synchronized to each other to generate optimal results.

7.3.4.2. Batch sizes (operated with)

For "fresh" raw materials: ~0.3 kgs to 700.0 kgs.
For "dry" raw materials: ~0.1 kgs to 275.0 kgs
For "wilted"/"semi-dryl" raw materials: anywhere in between.

However, batch sizes (with respect to fresh mint crop) of distillation "stills" in a proposed mixed plantation have been worked out (for MDPSystems) depending on the expected/available quantum of biomass from a project in a season or per annum as follows:

~375 kgs/batch (for a small-size mixed plantation up to *20 acres*);
~750 kgs/batch (for a medium-size mixed plantation up to *50 acres*); and
~2,000 kgs/batch (for a large-size mixed plantation up to *150 acres*), respectively.

MDPSystems can, however, be installed in a central location so as to work for a group of "entrepreneurs/farmers" in a particular locality (Karim, 2005).

7.4 RESULTS AND ANALYSIS

The purpose of distillation of essential oil crops or aromatic raw materials in pilot scale and commercial scale production is primarily to ensure the complete distillation of raw materials (at their different physical states: e.g., fresh, rain-fed, semi dry, dry and cured/matured etc.) for consistent and reproducible results in terms of yields and qualities in comparison to bench top or laboratory result and to achieve such results at the "shortest possible distillation-period" by conducting most of the distillation at about 1-atmospheric pressure (for superior quality yields/minimizing the chances of hydrolysis and decomposition of oil constituents by excessive contact-time with steam, heat and pressure) and at lowest distillation-cost preferably under some optimum regulated conditions. But effective hydro-distillation of different aromatic raw materials at the shortest distillation time and minimum cost poses a real challenge even today to scientists and engineers working in this field.

In the present studies, commercially complete hydro-distillation of *Menthol mint* (Mentha arvensis), *lemongrass* (Cymbopogon flexuosus), *citronella* (Cymbopogon winterianus), *Patchouli* (Pogostemon cablin) and *Vetiver* (Vetiveria zizanioides) etc. were repeatedly carried out in different batch-sizes and by the all three methods of hydro-distillation under required set of optimum-conditions in terms of post-harvest management (to maintain quality of raw materials), bulk-density of the charge, distillation temperature and pressure, steam quality, steam supply rate/rate of distillate etc. for batch to batch consistent/reproducible results with "precisionl" (in terms of yield and oil quality). Present results reveal that depending upon the prevailing physical state and actual essential oil content of the raw material variety, each type of raw material (i.e. each crop having different cultivation varieties, cvs.) possesses a respective (i.e. crop-specific) range of distillation-period per batch irrespective of feasible batch sizes which is found minimum for "the crop" in comparison to reported (published) and prevailing field-data of numerous "Conventional Distillation-Plants" at the grass-root levels in Indian Essential Oil Industry (IEOI) in particular. The achieved "minimum range of distillation period" per batch of a crop is termed here as the "Optimum Range of Distillation Period (ORDP)" for that particular crop; whereas the mean value of the said range is termed as the "Optimum Mean Distillation Period (OMDP)." The respective ORDP and CRDP ("Conventional Range of Distillation Period" in the prevailing Industry) and corresponding OMDP and CMDP ("Conventional Mean Distillation Period") of five types of raw materials, including mean-savings of time per batch, are presented in Table 7.1 (Karim, 2005, 2011) and Figure 7.1 (presenting comparison of respective "ORDP" versus "CRDP" etc.) respectively.

In hydro-distillation of all five types of raw materials the effect of steam-supply rate (alternatively, the effect of rate of "distillate") and other associated measures plays a major role in controlling/minimizing the processing time (i.e., duration of distillation) for distillation to take place under more "favorable atmosphere." Here, steam distillation of all five types of raw materials is graphically found to follow first-order reaction kinetics.

Basically, "hydro-distillation" is a two-step process, e.g., generation of steam in a separate vessel/boiler or within the distillation still (*in step 1*) and injection or flow of steam through the charged material for "distillation" to take place (*in step 2*). Moreover, the so called "distillation" takes place in 2-phases. In *phase 1* desired quality of steam at desired rate passes through the raw material to bring it to boiling temperature first and in the very next phase (or *phase 2*) distillation starts and mixed-vapor passes through the condenser, turned into liquids (i.e. water and oil), and finally gets separated. "Phase 1" and "phase 2" above are termed here as the "heating phase" of distillation and "distillation phase" of distillation respectively and both phases together constitute the "Period of distillation" per batch.

The oil-recovery patterns and the slope of progress of distillation of 4-aromatic crops, viz. Menthol mint crop, Lemongrass, Java-Citronella and Patchouli, in respective "distillation-phase" have been graphically presented (based on "single representative batch of distillation" for each crop) by plotting oil-yield % (along Y-axis) against time "t" in minutes (along X-axis) at different time-intervals (Figures 7.2 to 7.5).

Here, in four experiments above, length of "heating phases" and "distillation phases" of the crop materials under these experiments are respectively: 6.5 minutes and 53.5 minutes (for *Menthol mint*), 20 minutes and 60.5

Table 7.1 Crop-Specific Optimum Range and Conventional Range of Distillation-Period Per Batch of Five Raw Materials

		Range		*Mean*		
Sl. No.	*Raw material under trial*	*ORDP/batch (minutes)*	*CRDP/batch (minutes)*	*OMDP/ batch (minutes)*	*CMDP/ batch (minutes)*	*Mean savings/ batch (minutes)*
1	Menthol mint	51–69	180–360	60	270	210
2	Lemongrass	65–85	240–360	75	300	225
3	Citronella	105–135	210–480	120	345	225
4	Patchouli	160–250	360–1440	205	900	695
5	Vetiver	300–420	1440–1800	360	1620	1260

ORDP (Optimum Range of Distillation Period); CRDP (Conventional Range of Distillation Period); OMDP (Optimum Mean Distillation Period); CMDP (Conventional Mean Distillation Period)

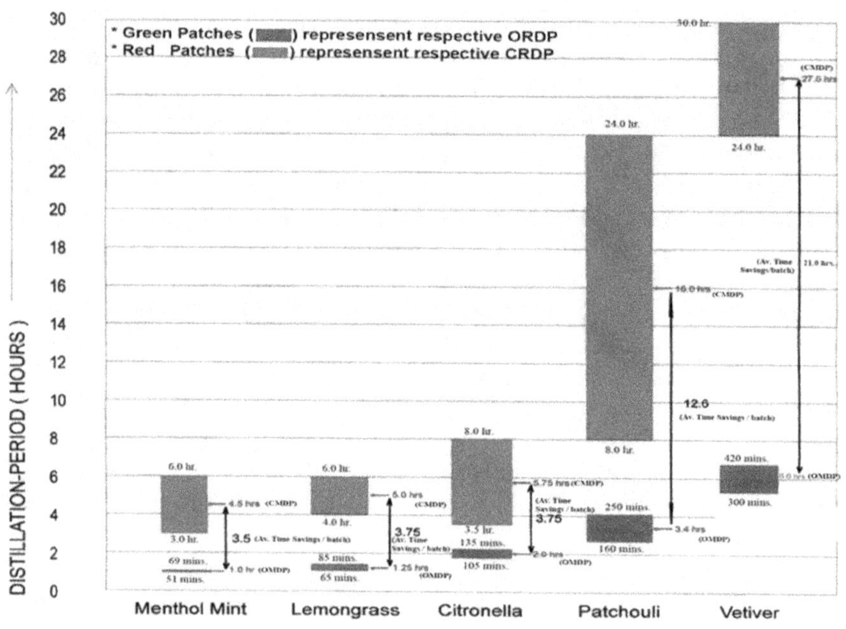

Figure 7.1 Comparison of ORDP versus CRDP of Respective Crops (as Presented in Table 7.1).

minutes (for *Lemongrass*), 29.5 minutes and 93.5 minutes (for *Java Citronella*) and 12 minutes and 209 minutes (for *Patchouli*).

Therefore "period of distillation" in these experiments are respectively: 6.5 + 53.5 = *60 minutes* (for Menthol mint), 20 + 60.5 = *80.5 minutes* (for

Lemongrass), 29.5 + 93.5 = *123 minutes* (for Java Citronella) and 12 + 209 = *221 minutes* (for Patchouli); where

Figure 7.2 represents the slope of Progress of Distillation of "Menthol mint crop" in its Distillation-Phase [in Yield% against Time (minute)];
Figure 7.3 represents the slope of Progress of Distillation of "Lemongrass" in its Distillation-Phase [in Yield% against Time (minute)];

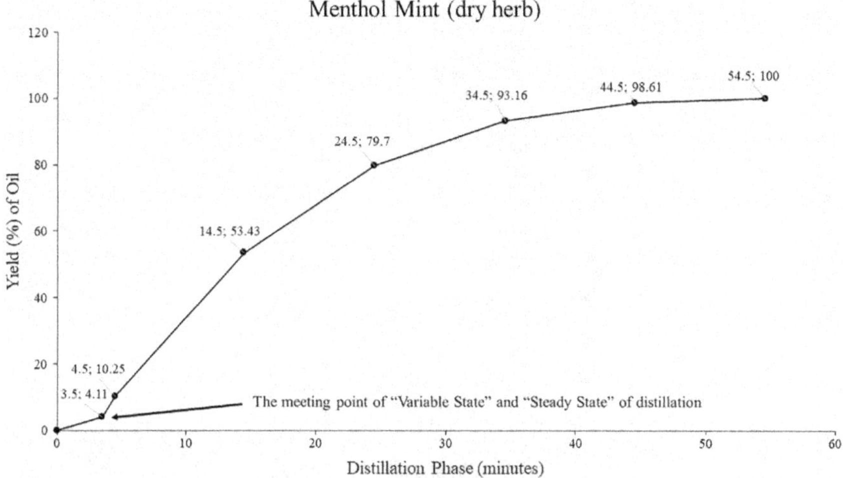

Figure 7.2 Progress of Distillation of Menthol Mint Crop.

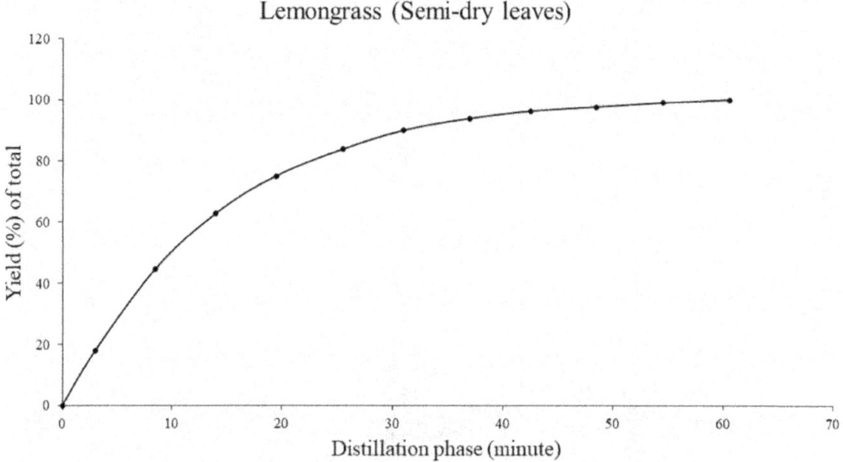

Figure 7.3 Progress of Distillation of Lemongrass Crop.

Figure 7.4 represents the slope of Progress of Distillation of "Java-Citronella" in its Distillation-Phase [in Yield % against Time (minute)]; and
Figure 7.5 represents the slope of Progress of Distillation of "Patchouli" in its Distillation-Phase [in Yield% against Time (minute)] respectively.

Moreover, in these experiments, bulk–density of load in *kg/L* of load volume and oil yield % (in v/W × 100) is respectively: 0.1188 kg/L and 0.7228%

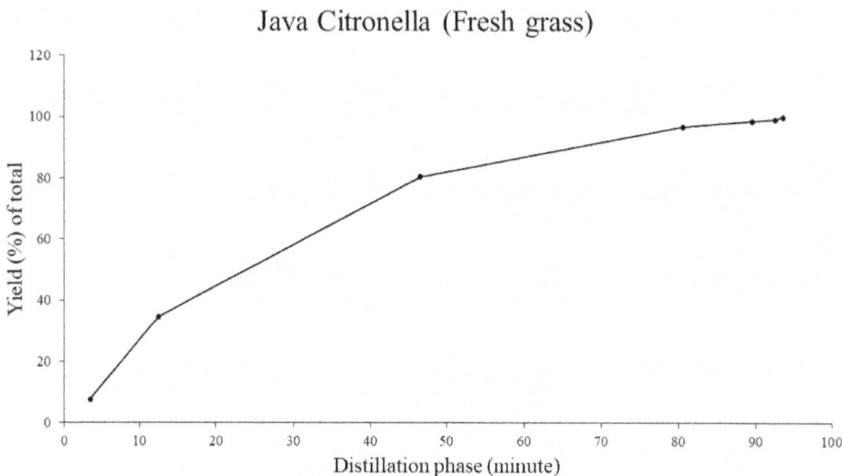

Figure 7.4 Progress of Distillation of Java Citronella Crop.

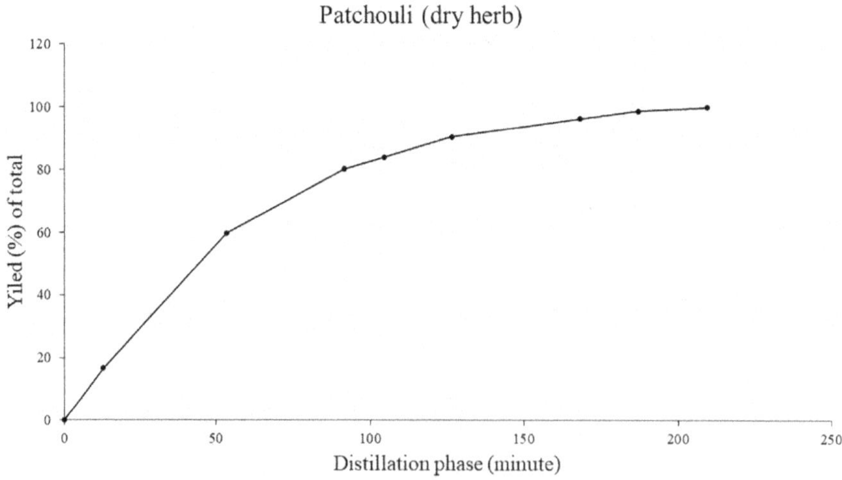

Figure 7.5 Progress of Distillation of Patchouli Crop.

(for *Menthol mint*), 0.251 kg/L and 0.836 % (for *Lemongrass*), 0.291 kg/L and 0.814% (for *Java-Citronella*) and 0.087 kg/L and 4.944% (for *Patchouli*). All such results qualify respective ORDP-Concept and other parameters of optimal distillation process.

Lastly, Figure 7.2 shows that for distillation of Menthol mint, "Variable State" of distillation phase and "Steady-State" of distillation phase meet together at 3.5 minutes with an yield of 4.09% oil (with respect to total oil yield); whereas Figures 7.2, 7.3, and 7.4 reflect only the "Steady State" of distillation phase.

Finally, the set-ups of MDP-System are made more environment friendly which allow remarkable savings on different overheads in comparison to conventional FDUs in IEOI; e.g. ~40.0% to 80.0% savings on time account, ~36.0% to 68.0% savings on steam account and ~34.0% to 64.0% savings on fuel account where the process-loss could be minimized to almost "zero" and ultimately the ORDP-Concept has emerged as a milestone for future "essential oil industry" which (ORDP-Concept) depicts the "minimum and sufficient distillation-time" per batch that a crop deserves for commercially complete distillation.

7.4.1 Process verification

In course of time, the so-called process of distillation of "Menthol mint crop" in the MDP-System at the village of Bandpur, Hooghly, was verified by Dr. Basab Choudhuri, then Reader, Department of Chemical Engineering, University of Calcutta, in 1999.

Similarly, the process of distillation of "citronella" and "patchouli" in the MDP-System of the Deganga Unit were verified in 2006 and certified by the Divisional Forest Officer, Minor Forest Produce Division, Siliguri, Darjeeling.

Lastly, the process of distillation of "Patchouli" in the 2nd Unit at Deganga village was verified and certified by the Manager, R&D and Projects, Ajmal Group, Flower Valley Agrotech Private Limited, Mumbai, in 2007.

7.5 DISCUSSION

7.5.1 Production and consumption rate

India has a vast essential oil industry. India's annual production of different essential oils is now (in *2022–2023*) estimated to be more or less 1,00,000 MT, of which the share of different mint oils is estimated to be around 65–70%, and, therefore, about 8.0–9.0 lakhs ha (hectares) of lands are estimated to be engaged under the cultivation of different aromatic plants for the production of required raw materials for the oils, estimated at ~20-million MT, little more or less. Moreover, based on modern market analysis tools, consumption of these items is also increasing at ~7.6 to 9.0 % annually during 2021–2027 (UnivDatos Market Insights Pvt. Ltd., 2022).

7.5.2 Financial outcomes

(i) Due to lack of post-harvest technology and management and numerous non-optimized distillation plants, the industry in India (IEOI) suffers from huge "process-loss" of valuable essential oils at ~16–20% annually, say, and it is estimated to be ~20,000–25,000 MT of *different oils* per annum, worth about 2,000–2,500 crores of rupees every year (this is the approximate loss incurred on the "lost oil" head considering average value of essential per kg at Rs.1,000.00); whereas Varshney S.C. in his report-book *Mint Oils of the World* assumed a process-loss of 10% (page 16).

(ii) Approximate loss incurred every year on the head of "distillation cost" (due to excess of fuel and man hours consumption than the optimum) = Rs. 800–1,000 crores.

(iii) Approximate additional loss incurred every year on the head of high "depreciation cost" of the conventional plants (due to short longevity and more numbers of plants) = Rs. 600–800 crores. Therefore, accumulated direct loss every year in the country due to the setup of improper plants = Rs. 3,400–4,300 crores; or approximately Rs. 4,000 crores annually. And, this huge amount of annual loss takes place out of negligence and/or ignorance/or unawareness of the people associated with the industry.

7.5.3 Feasible solution

Above accumulated amount of loss as well as major inadequacies of existing non-optimized DISTILLATION PLANTS in the industry could be avoided by modernization of the so-called inefficient PLANTS with the help of an "advanced distillation-technology" (as had been invented, optimized and now discussed in this chapter).

Most strikingly, one-time investment of such accumulated "lost amount" caused over a period of 2-years (i.e., more or less Rs.8,000 crores) is estimated to be enough to modernize the whole essential oil industry in India. Moreover, the payback period of invested money on such modernization head is hardly 2-years, rather 1–3 years more or less, depending on the extent of utilization of "enhanced-capacity" of so modernized units (out of much extra production and returns/income thereof). The avenues of earning continuous large revenues in subsequent years will thus be opened up. An in-depth survey and change of attitude are now essentially needed. In the long run, additional benefits that accrue would be more amount of quality products at low cost per unit time and higher rates of profit and better global competitiveness.

In a word, optimization of each individual UNIT of the industry, if carried out successfully, will lead to enhanced production and profit at less cost, thus causing improvement in the overall economy of this sector.

7.6 CONCLUSION

It is highly likely and desirable that, irrespective of feasible batch sizes, there should be a specific and "optimum range of distillation period (ORDP)" for each and every aromatic raw material. However, most of these measures are yet to be worked out carefully and implemented in the industry, as a whole. These will ensure the production of the maximum possible quantities of appropriate quality essential oils at the lowest cost per unit time in each and every unit of the industry. Obviously, such attempts are the need of the hour to cope with the present and future demands of the industry. In that case, no gap exists between the laboratory and industrial findings (i.e., lab-to-field gap would become zero in terms of various results). Moreover, this concept of ORDP is also commercially applicable to the "water distillation method" and the "water and steam distillation method" to a great extent, though with some reservations. In the above cases, the "boiling phase of water" and "heating phase" of distillation are merged together, and, therefore, the overall "heating phase" will be to some extent lengthy compared to the "direct steam distillation method."

It is the "essential oil industry" in general in India that has been suffering much since several decades to centuries, say, due to the use of mostly non-standardized conventional FDUs/other models in the large-scale production of different essential oils. Financially, the worst sufferers are the cultivators and the entrepreneur-distillers at the grassroot level; they are basically ignorant people, and a majority of them are not aware of the matter of so-called "process-loss" and "poor distillation efficiency" of their plants. This situation should be rectified in no time. It is true that the advanced distillation technology-based MDP-System would initially cost somewhat more than the conventional models, but in the long run the technology would win as it ensures several fold returns in terms of durability of the plant, a much higher rate of processing and production capacity per unit time (about 4 to 6 times), cost-effectiveness of superior and consistent quality products, and profit per unit production-mass per unit time. The sum of all descriptions is much higher than the rate of returns in the long run.

The MDP-System is easy to operate and environment-friendly. If centrally installed, it would work for the whole community of entrepreneurs of the locality.

One of the authors, Md. Fazlul Karim, has designed and operated such plants over a long period for a large number of aromatic crops and has demonstrated the efficiency of optimum distillation units. His expertise in this area may be utilized by those engaged in the essential oil industry.

Lastly, the authors firmly believe that there is enough scope for improvement in the essential oil industry, right from crop collection to final distillation, preferably by the use of "briquettes." This chapter is just an attempt toward convincing readers of the possibilities.

7.7 CONFLICT OF INTEREST

There is no conflict of interest.

ACKNOWLEDGMENT

The author is thankful to M. Masroor A. Khan, professor (plant pathology), Department of Botany, Aligarh Muslim University, Aligarh—202002 (UP), India, for sincerely making him aware of the proposed "book chapter" and also to Muhammad Farhan, Department of Mechanical Engineering, Faculty of Engineering, Integral University, Lucknow, for following up with the author at regular intervals on the advice of Professor Khan.

REFERENCES

Ahmed, M. 2002. Cultivars, Patchouli: An ideal aromatic crop of commercial importance; North Eastern Development Finance Corporation Ltd., pp. 21–22.

Ahmed, M. 2005. Patchouli: Hand book on medicinal & aromatic plants; NEDFI (North Eastern Development Finance Corporation Ltd.), pp. 1–9.

Akhila, A. and Tewari, R. 1984. Chemistry of patchouli oil, a review. *Current Research on Medicinal and Aromatic Plants* 6: 38–54.

Alkire, B.H. and Simon, J.E. 1990. Construction and design of a portable steam distillation unit for essential oil crops. *Hortscience* 25: 1165.

Anon ITC Report. 2014. Trade Impact for good—50 years (1964–2014): An overview of Mentha arvensis production; ITC Report, p. 1.

Bhan, M.K., Pal, S., Rao B.L. and Kaul B.L. 2001. Cost: Benefit analysis of lemongrass ("CKP-25") in Badaun district of Uttar Pradesh. *Indian Perfumer* 45: 173–177.

Bunrathep, S., Lockwood, G.B., Songsak, T. and Ruangrungsi, N. 2006. Chemical constituents from leaves and cell cultures of *Pogostemon cablin* and use of precursor feeding to improve patchouli alcohol level. *Journal Science Asia* 32: 293–296.

Cultivation of Patchouli—a manual from M/S. S. H. Kelkar & Co. Ltd. 2002. Lal Bahadur Shastri Marg, Mulund (West), Mumbai [https://keva.co.in].

Denny, E.F.K. 1990. Field distillation for herbaceous oils; Denny McKinzie Associates.

Dixit, S. 2019. Standardizing essential oils for purity and quality—a dire consumer need! *Indian Perfumer* 63: 17–27.

Dubey, K. 2022. Market report. *Indian Perfumer* 66(4): 72. ISSN0019-607X

Ecospice Ingredients Pvt. Ltd. 2023. Ecospice Farm, Thankamany (685609); Product/Farm bulletin [https://www.ecospiceingredients.com]

Farm Bulletin. 2004. Menthol mint cultivation. CIMAP (CSIR).

Farm Bulletin. 2007. Vetiver cultivation technology. CIMAP (CSIR), pp. 1–15.

Guenther, E. 1949. The essential oils. Vol. I; D Van Nostrand Company Inc., pp. 112–114, 121–122.

Imran, M., Liu, Y.H., Shafiq, S., Abbas, F., Ilahi, S., Rehman, N., Ahmar, S., Fiaz, S., Baran, N., Pan, S.G., Mo, Z.W., Tang, X.R. 2022. Transcriptional cascades in the regulation of 2-AP biosynthesis under Zn supply in fragrant rice. *Physiologia Plantarum*, 174 (3), Article e13721.

Jallan, A.K. 2004. Revival of essential oil industry in North Eastern India. Proceedings: AROMA [by North Eastern Development Finance Corporation Ltd.], p. 38.

Joy, P.P., Skaria, B.P., Mathew, S., Mathew, G. and Joseph, A. 2008. Status of aromatic crops in India. [Working paper, Kerala Agricultural University, Aromatic and Medicinal Plants Research Station, Odakkali, Asamannoor, Kerala], pp. 1–7. www.researchgate.net/publication/306015205

Karim, M.F. 2000. Agro-based "Vesaja Sugandhi" projects for self-employments. *Green Technology* 3: 20–32.

Karim, M.F. 2005. A concrete avenue towards globally competitive Indian essential oil industry. *Chemical Weekly*, 203–206.

Karim, M.F. 2006. Modern Distillation Plant (MDP) in preparation of agro based essential oil Industry in India; Section of Chemical Sciences, 93rd Indian Science Congress, p. 74.

Karim, M.F. 2011. Optimum-range of distillation period (ORDP) per batch of different aromatic raw-materials are generally different, minimum and crop-specific as well: an outstanding fact for industrial application; Book of abstracts, 3rd International Congress on Aromatic and Medicinal Plants (CIPAM-2011), p. 74.

Karim, M.F., Banerjee, S. and Poddar, M.K. 2018. Does patchouli oil change blood platelet monoamine oxidase-A activity of adult mammals? *The Journal of Physiological Sciences* 68: 281–291.

Kaul, V.K., Gandotra, B.M., Koul, S., Ghosh, S., Tikoo, C.L. and Gupta, A.K. 2004. Steam distillation of lemon grass (Cymbopogon spp.). *Indian Journal of Chemical Technology* (II) 135–139.

Lawrence, B.M. 1981. Patchouli oil: progress in essential oils. *Perfume Flavorist* 6: 73–76.

Lawrence, B.M. 1995. The isolation of aromatic materials from natural plant products, a manual on the essential oil industry (Ed. De Silva, K. T.). UNIDO.

Maheshwari, M.L., Vasantha, K.T. and Neelam, S., et al. 1993. Patchouli—an Indian Perspective. *Indian Perfumer* 37: 9–11.

Manglani, N., Deshmukh, V.S. and Kashyap, P. 2011. Evaluation of anti-depressant activity of pogostemon cablin (Labiatae). *International Journal of Pharm Tech Research* 3: 58–61.

Ozek, T. 2012. Distillation parameters for pilot plant production of Laurus nobilis essential *Oil Record of Natural Prod*ucts 6: 135–143.

Paul, D. 2013. Bio-resources: Potential source of biomass to utilize as alternative fuel for renewable energy; Proceedings, Section of Engineering Sciences, 100th Session of the Indian Science Congress, pp. 1–22.

Rajendra, P.Y., Alankararao, G.S.J.G. and Baby, P. 1987. Distillation of essential oils—certain inadequacies. *Indian Perfumer* 31: 1–6.

Ramya, H.G., Palanimuthu, V. and Rachna, S. 2013. An introduction to patchouli (Pogostemon cablin Benth.)—a medicinal and aromatic plant: it's importance to mankind. *Agricultural Engineering International CIGR Journal* 15: 243–250.

Sanganeria, S. 2005. Market overview-vibrant India—opportunities for the flavor and fragrance industry; Report: Ultra International Limited, pp. 1–12.

Satchith, K.M. and Sundaram, M. 1987. Production of essential oil and problems facing essential oil industry. *Indian Perfumer* 31: 41–43.

Singh, K. 1998. Production technology of Patchouli oil. In The cultivation and processing of medicinal and aromatic plants; Proceedings by CSIR—CIMAP at IICB, pp. 28–30.

Singh, S.P. and Singh, S. 1998. Production technology of Lemongrass oil. In The cultivation and processing of medicinal and aromatic plants; Proceedings by CSIR—CIMAP at IICB, pp. 10–13.

Tandon, S. 1998. Processing of medicinal and aromatic plants. In The cultivation and processing of medicinal and aromatic plants. Proceedings by CSIR—CIMAP at IICB, pp. 65–71.

TMV Aromatics Pvt. Ltd. 2023. TMV Gardens, Kochi 682035, Kerala, India: Product/ Farm bulletin [E-mail: info@tmvaroma.com; https://www.tmvaroma.com].

UnivDatos Market Insights Pvt. Ltd. 2022. Essential oils market is expected to display a steady growth by 2027 CAGR: ~9%. *Indian Perfumer* 66 (3): 12–13. ISSN0019-607X

Varshney, S.C. 1992. Essential oils by steam distillation (Ed. S.C. Varshney), 1st Edition, pp. 1–40.

Yahya, A. and Yunus, R.M. 2013. Influence of sample preparation and extraction time on chemical composition of steam distillation derived patchouli oil. *SciVerse Science Direct Procedia Engineering* 53: 1–6.

Chapter 8

Impact of biohumus on soil agrochemical characteristics, fertilty, and plant performance under greenhouse conditions

Khujamshukurov Nortoji A., Muhammad Farhan, Eshkobilov Sh.A., Kuchkarova D.X., Tufail Ahmad, Zainab Fatima, Sara Khan, and Juber Akhtar

8.1 INTRODUCTION

Due to the declining amount of arable land and the rising salinity of the soil and water, cucumbers have become a major crop in agriculture in recent years (Liu et al., 2020; Taha et al., 2020). Cucumis sativus L., the scientific name for cucumber, is a member of the Cucurbitaceae family and has 2n = 14 chromosomes (Obukhov and Efremova, 1988; Janapriya et al., 2010; Krug et al., 2002; Kubota et al., 2013). Cucumber is a widely grown crop in South Asia, especially in the hot and muggy parts of the Himalayas in northwest India and many North African nations. The climatic conditions in these areas are ideal for agricultural productivity, since they provide high temperatures, high humidity, plenty of sunshine, and easy access to water and nutrients. Cucumbers have a long and rich history that dates back to 3000 BC in India and 100 BC in China (Pal et al., 2020). The most popular crop grown in greenhouses worldwide is cucumber (Sharma et al., 2018; Singh, 2005; Bairagi et al., 2013; Egel, 2015; Fernandes et al., 2002). Cucumber growth and development under greenhouse conditions have been thoroughly investigated, with physiology and biochemistry as well as productivity-influencing elements being the subject of major scientific research (Sharma et al., 2018; Savvas et al., 2013). The effect of biohumus on the translocation of heavy metals during cucumber greenhouse culture and the mobility of nitrates within cucumber fruits, on the other hand, is conspicuously lacking in study in scientific literature (Khujamshukurov et al., 2022a, 2022b). Thus, the impetus of this study is to examine how biohumus affects the agrochemical characteristics of the soil, as well as its effects on fertility of the soil as well as production of cucumber under greenhouse conditions (Farhan, Mastura et al., 2022).

DOI: 10.1201/9781003495314-8

8.2 RESOURCES AND MATERIALS USED

This stage of the research sought to assess how organic fertilizer—more especially, biohumus—affects the substrate's agrochemical properties under carefully regulated soil conditions. The augmentation and advancement of the seeds of cucumber, specifically those of the "Avitsena" F1 cucumber variety, in a greenhouse setting was the main focus of the study. Established agricultural procedures that are often used in real-world applications were implemented in this study; the main differentiator was the experimental methodology (Farhan, Sadique et al., 2022). Two weeks before the substrate was moved into the planting beds, mineral fertilizers were added to it in these tests in accordance with the recommended procedure. The proper amount of biohumus was integrated before the substrate was added to the planting area. A rigorous procedure was used to prepare the cucumber seeds for planting, which included soaking them for a day in a diluted (1:50) Na3RO4 solution after an hour in a 1% KMnO4 solution under cold running water. Most of the seeds that had germinated by the time they were planted were put in 500 ml pots. Forty-five seedlings per square meter were sown, and until they germinated, they were covered with a clear polyethylene film. A constant temperature of 25–28°C was kept during the germination of the seedlings. The film was taken off after germination started, and the temperature was dropped to 22–24°C. Extra illumination was given at the seedling site for the first three days (a crucial step in situations of low winter solar radiation). After that, the seedlings were exposed to light for 12–16 hours per day at a light intensity of 4,000 and sometimes 5,000 lux. The air temperature was maintained at 19–20°C throughout the day's darkest hours. After 30 days, the seedlings were moved to their permanent locations in the greenhouse.

A 1:1 combination (soil: wood chips) was employed as a control. NH_4NO_3, Ca $(H_2PO_4)_2$, and K_2SO_4 were among the mineral fertilizers added to the control group's original combination to guarantee ideal nutritional levels. In the trials, 1 m^3 of the original mixture was mixed with 600 g of NH_4NO_3, 125 g of granulated Ca $(H_2PO_4)_2$, 455 g of KCl, and 115 g of $MgBr_2$ salts. Synergistic phenological observations, biometric investigations, and agrochemical analyses were used to evaluate the effect of biohumus on seedling growth (Dospekhov, 1985). The quality of a cucumber seedling significantly influences its subsequent output (Geisler et al., 1979). Seedlings should exhibit certain characteristics before planting, such as an average height of 20–25 cm, 4–5 real leaves, joint length of 3–5 cm, and a total weight ranging from 20 to 30 grams. Additionally, the veins should be disease-free and white (Farhan and Shafiq, 2019).

The content of nitrogen (A) and potassium (K) in one kg of dry greenhouse soil (mg) can be found by the following equations (Gluntsov et al., 1989): $A=(2\times O+15/10)/3$ (1). $K=(2\times O+15/20)/3$ (2).

Here, O is the amount of organic matter in the soil (in %).

8.3 OBTAINED RESULTS AND THEIR DISCUSSION

8.3.1 The influence of biohumus on soil chemical attributes and soil richness

The distinctive qualities of the cucumber plant are derived from its particular soil needs, which are fueled by its quick development and heavy nitrogen intake. It grows best on soil that has been sufficiently supplemented with a blend of mineral and organic fertilizers. Whereas phosphorus speeds up flowering and fruit production, nitrogen is essential for crop development and vegetative growth. Furthermore, potassium helps to raise the harvest's general quality. The following categorization system is used while growing cucumbers: Low supply is indicated by levels less than one-third of the needed amount; below the standard is indicated by levels between one-third (1/3) and two-thirds (2/3) of the required quantity; and levels at about two-thirds (2/3) meet the standard, quantities exceeding 1.33 times the requirement are considered above the norm, and significantly surpassing 1.33 times the requirement is classified as very high. Furthermore, when assessing phosphorus levels in all soil types during greenhouse cucumber cultivation, the criteria are measured in milligrams (mg) per 1 kilogram of completely dry soil.

According to the findings from the volumetric analysis, the data is outlined in Table 8.1

Cucumber plants in greenhouse environments require a steady supply of nutrients at every stage of their growth for a variety of reasons. First, for both stem and crop production, cucumbers need a constant intake of nutrients from the soil over a period of seven to nine months: 2.64 g of N_2, 1.55 g of P, 6.60 g of K, 2.19 g of Ca, and 0.57 g of Mg are needed to yield 1 kilogram of cucumbers. Twenty-three grams of nitrogen, 14 g of phosphorus, 58 g of potassium, 19 g of calcium, and 5 g of magnesium are found in one cucumber plant on average. Therefore, a nutrient-rich initial soil composition is required to sustain high yields throughout the growing

Table 8.1 The Soil Nutrient Provision Status in Greenhouses for Cucumber Farming

№	*Availability level*	*Amount of nutrients, mg/l*					*Electrical conductivity at 25∞C*
		N_{common}	*P_2O_5*	*K_2O*	*MgO*	*CaO*	
1	Low	0–30	0–3.5	0–38	0–18	0–36	0–0.5
2	Less than normal	31–60	3.6–7.0	39–76	10–36	37–72	0.6–1.0
3	Normal	61–90	7.1–10.5	77–115	37–54	73–108	1.1–2.0
4	High from normal	91–120	10.6–14.0	116–154	55–72	109–144	2.1–3.0
5	Very high	121–150	14.1–17.5	155–190	73–90	145–180	3.1–4.0

season. Cucumbers, however, cannot withstand extremely high nutritional concentrations in the soil, as it adversely affects the soil solution concentration and the soil's pH value. (The ideal pH range for cucumbers typically falls between 6.0 and 7.2.)

This is when the application of biohumus and the gradual addition of required nutrients becomes a sensible approach. Because cucumbers need a lot of water, using mineral fertilizers may cause some nutrients—most notably potassium and nitrogen—to seep out during irrigation. Furthermore, most nutrients—especially phosphorus—tend to bond with the soil and become less soluble. At different phases of growth, cucumber plants require varied amounts of nutrients.

Moreover, greenhouse soil's restricted capacity prevents it from holding on to the vital nutrients required to generate 30–40 kg/m^2 of cucumbers. Therefore, using phased or staggered feeding techniques to provide the required amount of fertilizer for growing greenhouse cucumber increases its efficacy (Ilyin, 1991).

To determine the fertilizer quantity applied both during soil preparation and subsequent feeding, the greenhouse soil's composition was analyzed before the experiment's commencement and on a monthly basis throughout the plant's entire development period. Agrochemical parameters for unfertilized greenhouse soil are detailed in Table 8.2.

The mean concentrations of potassium, phosphorus, and nitrogen in the soil of the unaltered plot over a three-year period (where neither organic nor mineral fertilizers were applied prior to plowing) are 280, 39, and 77 (in mg/kg), respectively, based on the data in Table 8.2. Nowadays, while growing cucumbers in greenhouses, there is more focus on making sure the soil has the right amount of macro- and micro-elements that are necessary for cucumber growth. Furthermore, a lot of focus is placed on adding physiologically active materials, including biohumus, that improve plant nutrition derived from the soil. Table 8.3 presents the results of the study for the important parameters of the biohumus and manure used in the studies (Imran et al., 2022).

Table 8.3 makes it clear that the amount of nutrients in biohumus fluctuates somewhat from year to year based on the makeup of the starting output, which is manure. However, in every case, the amounts of NPK (nitrogen, phosphorus, and potassium) in biohumus are significantly greater than the levels found in both manure and the soil within the experimental area. The agrochemical study, for example, showed that the produced biohumus had nitrogen, phosphorus, and potassium contents that were 5.2 times, 3.0 times, and 1.8 times greater, respectively, than those in manure. In addition, these values were greater than the soil in the experimental region by 2.6, 2.0, and 4.8 times, respectively. Because of its high nutritional content, biohumus can change the amount of nutrients in the soil. However, due to its inclusion of trace elements and humic acids, biohumus plays a role in

Table 8.2 Agrochemical Parameters of Unfertilized Greenhouse Soil

			Gross amount of NPK, %			*Amount of nutrients, mg/kg*							*Salt concentration, % soil to the mass relatively*
Years	*pH*	*Humus, %*	*N*	*P*	*K*	N_{com}	NO_3^-	NH_4^+	P_2O_5	K_2O	*CaO,%*	*MgO%*	
2019	7.2	2.64	0.23	0.26	2.2	74	57.5	16.5	39.5	280.0	4.8	0.6	0.18
2020	7.3	3.05	0.28	0.29	2.1	79	61.4	17.6	39.0	277.0	4.3	0.9	0.17
2021	7.4	2.86	0.24	0.26	2.0	78	61.1	16.9	38.5	283.0	4.1	0.9	0.19
3 years average	7.3	2.85	0.25	0.27	2.1	77	60.0	17.0	39.0	280.0	4.4	0.8	0.18

Table 8.3 Some Indicators of Manure and Biohumus Used in Greenhouse Conditions

№	*Determined indicator 3 years average*	*2019 years*	*2020 years*	*2021 years*	*3 years average*
Chemical composition of manure					
1	Moisture%	69.0	67.8	65.7	70.8
2	Organic substances%	21.8	23.3	26.9	24
3	N,%	0.53	0.48	0.45	0.49
4	P_2O_5, %	0.41	0.37	0.42	0.40
5	K_2O,%	0.62	0.80	0.96	0.79
6	CaO,%	1.47	1.35	1.34	1.39
7	MgO,%	0.17	0.14	0.12	0.14
Chemical composition of biohumus					
8	pH salty milk	7.3	7.0	7.3	7.2
9	Moisture %	62.0	60.6	60.8	61.1
10	Relative weight, g/cm3	0.71	0.73	0.72	0.72
11	Amount of humus, %	28.3	29.9	30.1	29.4
12	N, %	1.21	1.24	1.42	1.29
13	P_2O_5, %	0.83	0.77	0.83	0.81
14	K_2O,%	3.82	3.70	3.94	3.82
15	CaO,%	1.70	1.46	1.4	1.52
16	MgO,%	1.20	1.40	1.04	1.21
17	Cu, mg/kg	3.8	3.8	4.1	3.90
18	Zn, mg/kg	54.2	48.4	50.4	51.0
19	Co, mg/kg	0.3	0.4	0.2	0.3
20	Mn, mg/kg	61.8	64.0	63.8	63.2
21	Fe, mg/kg	258.0	277.0	266.0	267.0

enhancing cucumber plant development in greenhouse cultivation (Karpova, 2003).

Mineral fertilizers were applied to the greenhouse soil in the experimental control. The yearly application of mineral fertilizers was calculated. The plants were continuously given both organic and mineral fertilizers every 30 days in all cases.

Two phases involved the application of mineral fertilizers. In the first stage, the quantity of mineral fertilizers determined by the soil's chemical analysis was added. Varying approaches were considered when incorporating varying volumes of organic fertilizer, notably biohumus, in the second stage (Kasatikova, 2002).

In all cases, the same mineral fertilizer was used in these feedings: 29 mg/kg of nitrogen, 19.5 mg/kg of phosphorus (P_2O_5), and 23.5 mg/kg of potassium (as K_2O). NH_4NO_3 (22.0 kg/ha), double Ca $(H_2PO_4)2$ (25 kg/ha), and KCl (11 kg/ha) were the mineral fertilizers that were provided.

For agrochemical analysis, soil samples were taken both before and after each feeding. The soil's pH (rN value) and the amounts of nitrogen, phosphate, and potassium were measured employing well-proven agrochemical analytical techniques. The feeding schedule was followed as per the plan specified in the experimental area one to four times. Table 8.4 shows the features of the agrochemical greenhouse soil following the application of mineral and organic fertilizers prior to planting.

The aforementioned findings show that depending on the kind and amount of fertilizer used, the nutrient levels in the soil ready for seedling transplantation might differ significantly. Phosphorus varied from 60.7 to 117.0 mg/kg, potassium from 321.5 to 430.7 mg/kg, and nitrogen from 98.5 to 198.5 mg/kg. The soil in the control region, on the other hand, had a constant nutritional content of N-98.5 mg/kg, P-60.7 mg/kg, and K-321.5 mg/kg. These findings indicate that the field receives an ample supply of nutrients in accordance with well-recognized guidelines for cucumber production, even in the absence of biohumus. The consistent insertion of 1.0 kg/m^2 of biohumus that was tilled into the soil had positive results for the soil's physicochemical characteristics in every situation. The nutrient availability within the soil throughout the experiment varied depending on the nutrient content present in the biohumus used for each respective year. Throughout all the years, the nutrient levels in the control scenario consistently remained at or exceeded the ideal standards detailed in Appendix 8.1.

It was found that the pH value (rN) of the soil was not significantly affected by the addition of either manure or biohumus. The pH values of the

Table 8.4 Pre-planting Agrochemical Analysis of Greenhouse Soil Plowed with Different Amounts of Mineral and Organic Fertilizers (Biohumus, Manure) (Average 2019–2021)

			Amount of mobile NPK in the soil before transplanting, mg/kg		
№	*Options*	*pH*	*N*	*P_2O_5*	*K_2O*
1	$N_{350}P_{250}K_{170}$	6.67	98.5	60.7	321.5
2	$N_{350}P_{250}K_{170}$+$Б_{10}$(1)	6.73	120.8	74.4	344.2
3	$N_{350}P_{250}K_{170}$+$Б_{20}$(1+1)	6.90	124.8	75.1	335.0
4	$N_{350}P_{250}K_{170}$+$Б_{30}$(1+1+1)	6.83	125.1	80.0	332.5
5	$N_{350}P_{250}K_{170}$+$Б_{40}$(1+1+1+1)	6.90	123.2	75.0	330.2
6	$N_{350}P_{250}K_{170}$+$Б_{50}$) (1+1+1+1+1)	6.90	124.8	73.0	336.8
7	$Б_{50}$(2+1+1+1)	7.03	127.2	78.1	346.2
8	$Б_{80}$(3+2+1+1+1)	7.13	156.5	83.8	376.0
9	$N_{350}P_{250}K_{170}$+$Б_{40}$(40)	7.10	198.5	117.0	430.7
10	$N_{350}P_{250}K_{170}$+$Б_{90}$(3+2+2+2)	6.63	129.9	68.3	334.2

greenhouse soil in several situations where biohumus was employed ranged from 6.30 to 6.60, while the levels of salt content varied between 0.58% and 0.72%, roughly representing the ideal range. All of the varieties' soils had significantly higher levels of mobile N_2, K, and P after the addition of biohumus. This rise in nutritional content was shown in the first option that used biohumus. According to agrochemical assessments, biohumus raises the soil's nutrient levels to meet biohumus criteria (Table 8.4).

Based on the amount of accessible P2O5 in milligrams per kilogram of dry soil in a greenhouse, the phosphorus supply in the soil was assessed. Most plants can usually get enough phosphorus from soil that has more than 60 mg, and too much phosphorus does not harm a plant's ability to grow and develop (Gluntsov et al., 1989).

Table 8.4 shows that adding 1.0 kg/m^2 of biohumus is enough to enhance the phosphorus content in the soil by a substantial amount, with a mean increase of 12.2–15.0 mg/kg, or 17–20%, over the course of several years (as shown in Appendix 8.1). Nevertheless, a moderate addition of 1 kg/m^2 of biohumus resulted in just 3–5% rise in potassium concentration, rather than a significant change (Kutukova, 2001). The ninth option, which included 4.0 kg/m^2 of incorporated biohumus into the soil through plowing, had the maximum soil phosphorus accumulation (117.0 mg/kg). After comparing with controls, the total nitrogen content rose 1.30 times in the tenth choice (using 3.0 kg/m^2 of partially decomposed manure) and 1.60 times in the eighth option (using 3.0 kg/m^2 of biohumus). Additionally, the movable phosphorus content increased by 1.1 and 1.4 times, while potassium content increased by 1.1 and 1.2 times.

The nutritional content was 1.25 times greater in choices 2–5, where 1.0 kilogram of biohumus was injected per square meter of the experimental area, than in the control option. Levels of phosphorus were 1.10 and 1.23 times higher, respectively, and potassium was 1.25 times higher. Variants 7–9 demonstrated that, as indicated by Table 8.5, the amount of biohumus injected enhanced the nutritional content of the soil. The soil with the greatest average nutrient contents (N-195.8; P-117.0; K-430.7 mg/kg) was found in the 9th option, which included applying 4 kilogram of biohumus per square meter during plowing. Using 3 kilogram of partially decomposed manure per square meter beneath the plow, the 10th option's average NPK content was N-129.9, P-68.3, and K-334.2 mg/kg, closely aligning with nutrient levels in alternatives 2–6, all involving the application of 1.0 kg/m^2 of biohumus (Kutukova and Plekhanova, 2002; El-Gamal et al., 2016).

The trials showed that the plant had already absorbed a sizable amount of the total nutrients from the soil during the time leading up to the first feeding, or before blossoming. In particular, around 22–27% of the total N_2, 19–23% of the P, and 29–33% of the K had been absorbed by the plant. As shown in Appendix 8.2, the total nutrient absorption coefficient varied

Table 8.5 The Result of Pre-harvest Agrochemical Analysis of Greenhouse Soil (Average Amount of NPK, 2019–2021, mg/kg)

№	Options	pH	NPK available in greenhouse soil before first feeding, mg/kg		
			N	P_2O_5	K_2O
1	$N_{350}P_{250}K_{170}$	6.70	72.2	43.7	210.2
2	$N_{350}P_{250}K_{170}$+$Б_{10}$(1)	6.77	92.1	56.0	233.2
3	$N_{350}P_{250}K_{170}$+$Б_{20}$(1+1)	6.87	88.1	59.8	230.1
4	$N_{350}P_{250}K_{170}$+$Б_{30}$(1+1+1)	6.83	90.5	56.3	224.9
5	$N_{350}P_{250}K_{170}$+$Б_{40}$(1+1+1+1)	6.90	86.9	60.5	230.6
6	$N_{350}P_{250}K_{170}$+$Б_{50}$) (1+1+1+1+1)	6.90	95.0	55.2	232.8
7	$Б_{50}$(2+1+1+1)	7.00	89.4	54.2	243.1
8	$Б_{80}$(3+2+1+1+1)	7.10	116.0	63.6	260.5
9	$N_{350}P_{250}K_{170}$+$Б_{40}$(40)	7.00	148.8	101.8	274.9
10	$N_{350}P_{250}K_{170}$+$Б_{90}$(3+2+2+2)	6.73	102.2	54.5	218.4

depending on the scenario, ranging from 1.25 to 1.35 for nitrogen, 1.25 to 1.30 for phosphorus, and 1.42 to 1.56 for potassium.

During the course of the tests, a feeding plan incorporating one to four fertilization cycles was put into place. Table 8.6 provides specifics based on the outcomes of the soil analysis performed following the first feeding, which corresponds to the plant's growth stage (Obukhov and Efremova, 1988).

Following the initial feeding, there is a considerable increase in nutrient content in all situations, according to the findings of the soil composition study shown in Table 8.6. In variations 3–9, when biohumus-based fertilization was used, the rise was very noticeable, with the NPK content of the soil increasing in direct proportion to the amount of biohumus applied. According to Appendix 8.3, the increase in NPK content relative to the control option varied from 1.41 to 1.70 times for nitrogen, 1.47 to 1.68 times for phosphorus, and 1.17 to 1.22 times for potassium in all scenarios utilizing 1.0 kg/m^2 of biohumus.

Tables 8.7 and 8.8 display the findings of soil analysis in the greenhouse, both before feeding (Table 8.7, as presented in Appendix 8.4) and after feeding (Table 8.8, as provided in Appendix 8.5). These analyses were conducted during the period of cucumbers' full ripening.

Table 8.7 shows that throughout the main crop's ripening period, there is a significant drop in nutrient levels in every scenario. Notably, at 35.4 mg/kg, the control variety had the lowest nitrogen concentration prior to the second meal. On the other hand, the nitrogen content of the soil in the control scenario rose to 66.1 mg/kg by applying mineral fertilizers alone (Table 8.8).

Table 8.6 The Quantity of Nutrients Collected in Greenhouse Soil Following the Initial Fertilization (Average from 2019 to 2021)

№	Options	pH	*Amount of total NPK accumulated in greenhouse soil after first feeding, mg/kg*		
			N	P_2O_5	K_2O
1	$N_{350}P_{250}K_{170}$	6.80	101.2	66.0	233.2
2	$N_{350}P_{250}K_{170}$+$Б_{10}$(1)	6.80	138.1	87.0	281.1
3	$N_{350}P_{250}K_{170}$+$Б_{20}$(1+1)	6.87	156.2	98.2	296.1
4	$N_{350}P_{250}K_{170}$+$Б_{30}$(1+1+1)	6.97	158.3	97.6	294.4
5	$N_{350}P_{250}K_{170}$+$Б_{40}$(1+1+1+1)	6.90	165.9	100.6	295.6
6	$N_{350}P_{250}K_{170}$+$Б_{50}$) (1+1+1+1+1)	6.90	162.2	110.1	299.6
7	$Б_{50}$(2+1+1+1)	7.10	154.7	87.1	312.0
8	$Б_{80}$(3+2+1+1+1)	7.10	215.2	125.0	378.3
9	$N_{350}P_{250}K_{170}$+$Б_{40}$(40)	7.07	241.3	151.8	397.8
10	$N_{350}P_{250}K_{170}$+$Б_{90}$(3+2+2+2)	6.53	164.4	84.8	292.5

Table 8.7 The NPK content present in greenhouse soil before the second plant feeding during the period of high productivity (average data from 2019 to 2021)

№	Options	pH	*NPK content in greenhouse soil before second feeding, mg/kg*		
			N	P_2O_5	K_2O
1	$N_{350}P_{250}K_{170}$	6.80	35.4	32.5	113.4
2	$N_{350}P_{250}K_{170}$+$Б_{10}$(1)	6.67	59.7	49.0	152.2
3	$N_{350}P_{250}K_{170}$+$Б_{20}$(1+1)	6.83	77.5	55.5	160.6
4	$N_{350}P_{250}K_{170}$+$Б_{30}$(1+1+1)	6.80	79.7	57.5	165.2
5	$N_{350}P_{250}K_{170}$+$Б_{40}$(1+1+1+1)	6.90	83.3	57.1	168.0
6	$N_{350}P_{250}K_{170}$+$Б_{50}$)(1+1+1+1+1)	6.90	79.1	66.2	174.5
7	$Б_{50}$(2+1+1+1)	7.00	71.3	42.6	167.1
8	$Б_{80}$(3+2+1+1+1)	7.07	121.7	75.6	237.5
9	$N_{350}P_{250}K_{170}$+$Б_{40}$(40)	6.97	163.0	104.0	230.0
10	$N_{350}P_{250}K_{170}$+$Б_{90}$(3+2+2+2)	6.67	82.1	36.1	142.4

The soil of the second through seventh varieties had a nitrogen concentration ranging from 60 to 80 mg/kg throughout the ripening period preceding the second feeding. When the cucumber crop reached full ripening, there was a noticeable drop in the amount of phosphorus present, especially in choices 1, 7, and 10, where manure was applied. In these circumstances, the lowest phosphorus levels ranged from 32 to 42 mg/kg.

The soil's nutrient levels were greater in 2021 when the results of soil analyses for the same area and options were compared between 2020 and

Tables 8.7–8.8 show the results of greenhouse soil analysis before feeding (Table 8.7, Appendix 8.4) and after feeding (Table 8.8, Appendix 8.5) carried out by the period of gross ripening of cucumbers

№	Options	pH	*Amount of NPK after the second feeding, mg/kg*		
			N	P_2O_5	K_2O
1	$N_{350}P_{250}K_{170}$	6.73	66.1	51.7	150.0
2	$N_{350}P_{250}K_{170}$+Б$_{10}$(1)	6.83	105.2	75.1	195.9
3	$N_{350}P_{250}K_{170}$+Б$_{20}$(1+1)	6.90	136.3	89.4	229.1
4	$N_{350}P_{250}K_{170}$+Б$_{30}$(1+1+1)	6.80	149.5	109.6	260.5
5	$N_{350}P_{250}K_{170}$+Б$_{40}$(1+1+1+1)	6.97	162.5	105.7	257.7
6	$N_{350}P_{250}K_{170}$+Б$_{50}$) (1+1+1+1+1)	6.93	155.3	113.2	265.7
7	Б$_{50}$(2+1+1+1)	7.00	122.1	78.7	248.1
8	Б$_{80}$(3+2+1+1+1)	7.10	232.6	129.0	385.3
9	$N_{350}P_{250}K_{170}$+Б$_{40}$(40)	6.97	213.4	140.8	327.6
10	$N_{350}P_{250}K_{170}$+Б$_{90}$(3+2+2+2)	6.60	155.1	70.8	229.8

2021, the major crop development season. Tables 8.7 and 8.8 show that in the month preceding the primary harvest's maturity, the potassium level of the soil dropped by 1.6–1.8 times. Options 4–6, which used biohumus, showed a notable increase in the number of nutrients in the soil affects the second feeding (as reported in Appendix 8.5).

In the eighth scenario where biohumus was applied, the highest recorded nitrogen level in the soil reached 232.6 mg/kg, whereas the lowest was observed at 149.5 mg/kg in the second scenario (Alekseev, 1987). With the exception of choices 1–3, where only mineral fertilizers were given, potassium content varied from 248–385 mg/kg and phosphorus levels decreased between 105 and 140 mg/kg after the second feeding. As a result, during the growing phase, all scenarios showed a reasonably high nutritional content.

The 3rd and 4th cucumber plant feedings were conducted as planned, aligning with the maturation of the initial crop. The findings of soil composition studies conducted in the experimental field, both before and after the plant was fed, for various third-stage feeding scenarios are shown in Tables 8.9 and 8.10.

Upon reviewing these data, it can be observed that the plant absorbed 39.8–70.7% of the nitrogen, 23.4–52.1% of the phosphorus, and 39.51–59.8% of the potassium across several alternatives during the period of peak productivity that preceded the third feeding. choices 1 and 3 for N_2, choices 1 and 7 for P, and options 1 and 5 for K all showed a discernible decline in nutrient levels.

Among the benefits of using biohumus instead of mineral fertilizers for growing cucumbers in a greenhouse is that the former may hold onto nutrients in its composition rather of releasing them into the soil solution right away (Obukhov and Plekhanova, 1995). Rather, it releases nutrients in the amounts that the plant needs. This characteristic is exemplified in Appendix 8.6 and Table 8.9, you can find the outcomes of the soil examination performed before the third feeding.

There is a noticeable difference in the NPK level of the soil in choices 8 and 9, where biohumus was the only fertilizer utilized, and the control option, which only employed mineral fertilizers. Unlike the choices that included biohumus, the control option with mineral fertilizers saw a considerable decline in the nutritional components in the time leading up to the next fertilization. When biohumus was used in variations 8 and 9, the plant continuously produced substantial amounts of NPK during the course of its development.

Table 8.9 displays the NPK levels in the greenhouse soil before the third feeding, recorded during the period of robust plant growth, with data averaged over the years 2019 to 2021.

By this stage of the experiment, it had become apparent that in greenhouse soil selections 5–8, where biohumus was included in the fertilization routine, nutrient levels were notably higher than in those alternatives which solely relied on mineral fertilizers without the addition of biohumus. After the third round of fertilization, a thorough analysis of Appendix 8.7 and Table 8.10 revealed that choices 5–8, which utilized both biohumus and mineral fertilizers, experienced a significant increase in nitrogen content, ranging from 20 to 65 mg/kg or a 16 to 82% increase, phosphorus levels increased by 20 to 50 mg/kg or 20 to 71%, and potassium levels by 43 to 84 mg/kg, resulting in a rise of 26 to 67% when compared to options 1–4, where only mineral fertilizers were employed. A similar trend was observed

Table 8.9 The Amount of NPK in the Soil of the Greenhouse Before the Third Feeding During the Period of Gross Plant Production (Average, 2019–2021)

			NPK content in greenhouse soil before the third feeding, mg/kg		
№	*Options*	*pH*	*N*	*P_2O_5*	*K_2O*
1	$N_{350}P_{250}K_{170}$	6.80	19.7	25.0	63.7
2	$N_{350}P_{250}K_{170}$+Б$_{10}$(1)	6.83	41.4	45.0	116.6
3	$N_{350}P_{250}K_{170}$+Б$_{20}$(1+1)	6.87	39.9	47.8	120.0
4	$N_{350}P_{250}K_{170}$+Б$_{30}$(1+1+1)	6.80	69.9	58.7	107.1
5	$N_{350}P_{250}K_{170}$+Б$_{40}$(1+1+1+1)	6.93	68.0	58.3	103.6
6	$N_{350}P_{250}K_{170}$+Б$_{50}$) (1+1+1+1+1)	6.90	64.5	69.6	114.8
7	Б$_{50}$(2+1+1+1)	6.97	39.8	37.7	104.7
8	Б$_{80}$(3+2+1+1+1)	7.10	140.0	98.8	233.2
9	$N_{350}P_{250}K_{170}$+Б$_{40}$(40)	7.00	112.0	96.9	168.2
10	$N_{350}P_{250}K_{170}$+Б$_{90}$(3+2+2+2)	6.60	82.4	34.2	95.1

Table 8.10 The Total Accumulation of NPK in the Soil after the Third Feeding, Throughout the Period of Vigorous Plant Growth (Averaged over the Years 2019–2021)

№	*Options*	*pH*	*Amount of NPK in the soil after the third feeding, mg/kg*		
			N	*P_2O_5*	*K_2O*
1	$N_{350}P_{250}K_{170}$	6.73	42.0	44.4	83.6
2	$N_{350}P_{250}K_{170}$+$Б_{10}$(1)	6.77	79.6	66.9	110.1
3	$N_{350}P_{250}K_{170}$+$Б_{20}$(1+1)	6.80	82.2	70.2	123.9
4	$N_{350}P_{250}K_{170}$+$Б_{30}$(1+1+1)	6.83	120.5	99.8	170.2
5	$N_{350}P_{250}K_{170}$+$Б_{40}$(1+1+1+1)	6.87	147.1	112.5	186.6
6	$N_{350}P_{250}K_{170}$+$Б_{50}$) (1+1+1+1+1)	6.90	143.6	121.0	207.7
7	$Б_{50}$(2+1+1+1)	6.93	87.0	68.7	184.5
8	$Б_{80}$(3+2+1+1+1)	7.07	210.5	144.0	324.5
9	$N_{350}P_{250}K_{170}$+$Б_{40}$(40)	6.93	145.6	122.6	198.6
10	$N_{350}P_{250}K_{170}$+$Б_{90}$(3+2+2+2)	6.63	165.4	68.1	177.0

in the selection using manure, as each round of feeding led to increased levels of N_2, P, and K when contrasted with the control choice.

The soil in the greenhouse associated with choices 1–4, where biohumus was added up to twice throughout the experiment, showed increases in NPK content in direct proportion to the number of biohumus feedings. For example, the nitrogen level in version 2, where biohumus was administered once, was 61% higher than that of the control variant, which did not apply biohumus. The soil in option 4, where the treatment of biohumus was discontinued after two feedings, nevertheless, had a high level of nitrogen (120.5 mg/kg).

Similar patterns can be observed for potassium and phosphorus as well. After the third feeding, the phosphorus and potassium levels in options 5 and 6, where both mineral fertilizer and biohumus were applied, exhibit the same trends, were higher than those of option 4, which had only mineral fertilizer applied in the third feeding and had used biohumus for the first two feedings. The differences were 12–20 mg/kg and 22–43 mg/kg, respectively.

It was found that in the third year of the trial, the soil of the biohumus-treated option had a greater nutritional content than in the first. For instance, option 5 showed NPK levels of 129.9 mg/kg, 91.5 mg/kg, and 165.7 mg/kg in the first year following the third feeding, and in the third year of the experiment, these levels had increased to N-165.6 mg/kg, P-126.6 mg/kg, and K-205.8 mg/kg (as indicated in Appendix 8.7).

When compared to the choices that employed biohumus, the increase in soil nutrient content was less noticeable even though the dung-based options used a larger quantity of manure. When biohumus feeding was stopped during the cucumber's gross production period, the soil analysis in those choices showed a notable decline in soil nutrient levels in the later stages of plant growth. In the control variation, N_2 fell from 42.0 mg/kg to 16.2 mg/kg, P from 44.4 mg/kg to 25.4 mg/kg, and K from 83.6 mg/kg to 20.6 mg/kg between the third and fourth feedings. The overall NPK content in choices 1–3 by this point (before the fourth feeding) was, on average, 25–40% of the total NPK that had accumulated following the third feeding.

After the third application of fertilizers, the first control option had the highest nitrogen absorption rate at 61%, while the eighth option had the lowest absorption rate at 27%. The total phosphorus absorption by plants across different options varied from 25% to 43%, and potassium absorption ranged from 31% to 75%. According to the experimental plan, the fourth fertilization round involving both mineral fertilizers and biohumus was implemented only in option 6.

Table 8.12 shows that the soil of option 6 has much higher amounts of potassium, phosphorus, and mobile nitrogen. The levels of these nutrients have grown by 1.4, 2, and 2.4 times, respectively, compared to their original levels prior to the feeding in this option, when biohumus was applied at a rate of 1 kg/m² during the fourth feeding (Table 8.11 and Appendix 8.8).

The amount of nutrients in the soil increased significantly as a result of repeated feedings with biohumus. Options 1–3 showed significantly decreased NPK concentration when biohumus was excluded from future feedings. By harvest time, the NPK level of the soil in the experimental field had significantly decreased during the course of the research. At this point, nevertheless, the biohumus-fed variations continued to have an advantage over the control type in terms of NPK levels. The amount of mineral

Table 8.11 The NPK Composition Within the Greenhouse Soil Prior to the Fourth Plant Feeding, Recorded During the Period of Vigorous Plant Growth and Averaged over the Years 2019–2021

			Amount of NPK in the soil before fourth feeding, mg/kg		
№	*Options*	*pH*	*N*	*P_2O_5*	*K_2O*
1	$N_{350}P_{250}K_{170}$	6.73	16.2	25.4	20.6
2	$N_{350}P_{250}K_{170}$+$Б_{10}$(1)	6.73	33.4	49.9	58.3
3	$N_{350}P_{250}K_{170}$+$Б_{20}$(1+1)	6.77	34.8	45.3	43.8
4	$N_{350}P_{250}K_{170}$+$Б_{30}$(1+1+1)	6.80	60.3	65.2	55.8
5	$N_{350}P_{250}K_{170}$+$Б_{40}$(1+1+1+1)	6.87	58.7	69.1	52.9
6	$N_{350}P_{250}K_{170}$+$Б_{50}$) (1+1+1+1+1)	6.87	62.2	86.2	64.6
7	$Б_{50}$(2+1+1+1)	6.90	40.6	43.0	88.7
8	$Б_{80}$(3+2+1+1+1)	6.97	153.7	115.3	223.8
9	$N_{350}P_{250}K_{170}$+$Б_{40}$(40)	6.97	98.7	95.0	108.8
10	$N_{350}P_{250}K_{170}$+H_{90}(3+2+2+2)	6.63	97.7	36.3	68.9

Table 8.12 The NPK Levels in the Soil Following the Fourth Feeding, Taking Place During the Period of Abundant Plant Yield and Calculated as an Average from 2019 to 2021

№	*Options*	*pH*	*NPK content in greenhouse soil after 4-feeding, mg/kg*		
			N	*P_2O_5*	*K_2O*
1	$N_{350}P_{250}K_{170}$	6.70	45.3	46.1	43.2
2	$N_{350}P_{250}K_{170}$+Б$_{10}$(1)	6.70	63.4	66.3	81.6
3	$N_{350}P_{250}K_{170}$+Б$_{20}$(1+1)	6.80	58.8	66.3	70.6
4	$N_{350}P_{250}K_{170}$+Б$_{30}$(1+1+1)	6.80	89.9	91.1	92.9
5	$N_{350}P_{250}K_{170}$+Б$_{40}$(1+1+1+1)	6.90	107.7	103.3	99.7
6	$N_{350}P_{250}K_{170}$+Б$_{50}$) (1+1+1+1+1)	6.90	123.7	122.5	157.6
7	Б$_{50}$(2+1+1+1)	7.00	65.6	62.7	133.6
8	Б$_{80}$(3+2+1+1+1)	7.10	188.9	137.3	267.1
9	$N_{350}P_{250}K_{170}$+Б$_{40}$(40)	7.00	126.4	114.3	148.8
10	$N_{350}P_{250}K_{170}$+Б$_{90}$(3+2+2+2)	6.67	127.0	59.3	108.3

nitrogen in the soil rose in direct proportion to the quantity of feedings in all cases. Although there were few occasions where the N_2 content determination findings under the same option differed across years (Appendix 8.9), these changes were not statistically significant in the majority of cases. For both potassium and phosphorus, a comparable pattern was seen.

The amount of potassium in the soil is essential for increasing plant output and product quality while growing cucumbers in a greenhouse. When producing cucumbers, a potassium deficit in the soil typically results in a higher percentage of subpar yields. Following harvest, there was a constant trend in the K_2O levels of the soil during the whole research period. Significant biohumus consumption kept a comparatively high quantity of exchangeable potassium in the variants.

All things considered, the findings of agrochemical soil analysis carried out at various phases of plant growth during experiments on the growing of cucumbers in greenhouses showed a notable decline in soil nutrient levels, which corresponded to the buildup of biomass by harvest time. Notably, the use of mineral fertilizers along with biohumus in greenhouse cucumber cultivation helped to maintain soil nutrient levels, particularly nitrogen, at levels similar to those before seedling planting throughout various plant development stages. This trend was also evident in the levels of P_2O_5 and K_2O in the greenhouse soil post-harvest. Over the three-year research period, soil nutrient content in the plow layer of the experimental field was monitored after the conclusion of the experiment. Table 8.13 presents the three-year average of soil nutrient levels following the completion of the experiment from 2019 to 2021.

Table 8.13 The Findings from the Examination of Soil Composition Following the Complete Harvest (Averaged over the Years 2019–2021)

№	*Options*	*pH*	*Amount of NPK in the soil at the end of the growing season, mg/kg*		
			N	P_2O_5	K_2O
1	$N_{350}P_{250}K_{170}$	6.73	18.9	32.8	26.6
2	$N_{350}P_{250}K_{170}$+$Б_{10}$(1)	6.77	27.9	51.7	49.1
3	$N_{350}P_{250}K_{170}$+$Б_{20}$(1+1)	6.87	34.9	49.8	29.0
4	$N_{350}P_{250}K_{170}$+$Б_{30}$(1+1+1)	6.80	45.5	73.8	38.1
5	$N_{350}P_{250}K_{170}$+$Б_{40}$(1+1+1+1)	6.90	50.8	80.3	50.6
6	$N_{350}P_{250}K_{170}$+$Б_{50}$) (1+1+1+1+1)	6.93	64.0	101.4	100.4
7	$Б_{50}$(2+1+1+1)	6.97	27.9	45.2	92.9
8	$Б_{80}$(3+2+1+1+1)	7.03	110.0	119.3	221.0
9	$N_{350}P_{250}K_{170}$+$Б_{40}$(40)	7.00	93.0	101.1	92.5
10	$N_{350}P_{250}K_{170}$+H_{90}(3+2+2+2)	6.77	91.0	48.7	69.0

During the third year of the greenhouse experiment, the residual nutrient levels (NRC) in the soil exhibited a substantial increase, ranging from 1.8 to 2.0 times higher when compared to the initial levels recorded in the experiment's first year, as indicated by the analysis of post-harvest soil results (refer to Appendix 8.10). Notably, there was no similar elevation in nutrient levels in the later years of the experiment for the control group, which did not utilize biohumus.

There is an increase in nutritional content from 2019 to 2021 when the post-feeding NPK levels and the pre-feeding NPK levels for each option are compared. In particular, there was an increase in potassium (K) of 1.2 to 2.3 times, phosphorus (P) of 1.2 to 2.0 times, and nitrogen (N) of around 1.2 to 1.7 times. This rise was most noticeable in choices 5 through 8, when biohumus was administered in phases at a rate of 1.0 kg/m^2. For example, Option 5 showed increases in soil nitrogen (50 mg/kg) and phosphorus (53 mg/kg) and potassium (49 mg/kg) to 90 mg/kg and 101 mg/kg, respectively, in the third year.

By the time the experiment was over, the NPK levels in the experimental soil under option 9, which involved adding 4.0 kg/m^2 of biohumus to the soil before planting, had dropped below the ideal range. In contrast, at the end of the trial, there was a noticeable drop in the NPK content in the region where 2 kg/m^2 of manure was treated for fertilization. This demonstrates how the dynamics of nutrients vary under several experimental settings.

The level of vital minerals needed for plant development, including micro and macroelements in their mobile form, was found to be significantly diminished in the soil of the control group. This reduction resulted from the intensive use of the soil within greenhouse conditions, where only mineral fertilizers were employed. To achieve a high yield under the circumstances of closed soil conditions, it is often necessary to utilize high-quality mineral fertilizers and other chemical compounds. Consequently, the crop begins to amass an excessive number of toxic compounds. Preventing uncomfortable conditions during feeding is achieved by using biohumus (Plekhanova et al., 1995a; Plekhanova and Savelyeva, 1997).

8.3.2 The effect of biogumus feeding on the dynamics of change of N-NO_3 and N-NH_4 in the soil

When cucumbers are grown under greenhouse conditions, the optimal amount of mineral nitrogen present in the soil is 61–90 mg/l, and the plant absorbs nitrogen from the soil in the form of NH_4+-cation and NO_3-anion. Generally, the amount of nitrogen in the nitrate form in the soil is greater than its amount in the ammonium form, and it is the most absorbable form of nitrogen by the plant.

In order to reduce the amount of nitrates in the soil during the initial period of the formation of the cucumber crop and thereby reduce the formation of nitrates in the crop, it is advisable to use the organic fertilizer biohumus, which separates the nutrients contained in the soil solution in accordance with the absorption of the plant and has a prolonging property, along with mineral fertilizers.

In the experiment, it was observed that in the options where biohumus was used, the amount of nitrates in the soil solution in the initial period of vegetation (the first month) was lower than in the options where an equivalent amount of mineral fertilizer was used. Therefore, the use of biohumus allows a reduction of accumulation of nitrates in the plant and its crop and an improvement in the quality of the product.

Table 8.14 displays the outcomes of nitrate level determination in greenhouse soil that was prepared for planting seedlings employing various methods and following that was subjected to the second and fourth feedings.

According to the data presented in Table 8.14, the changes in nitrate content within the soil of the experimental field, across different years of the experiment, remained consistent in variants that received biohumus. The most significant reduction in N-NO_3 content, during the period from planting to the second and fourth fertilization, was observed in the control option. In this scenario, the total N-NO_3 aggregated in the soil after the second feeding was 75% of the initial N-NO_3 content in the area before planting. After the fourth feeding, this figure decreased to 50%. In contrast, in option 2, where 1 kg/m^2 of biohumus was plowed into the soil, these values were 85% and 56%, respectively. It's noteworthy that, in all the variants where biohumus was used for fertilization, unlike the control option, the N-NO_3 levels in the soil remained stable.

Table 8.14 Fluctuations in the Quantity of N-NO_3 Within the Experimental Field Soil Following the Application of Biohumus With Varying Specifications (Post-feeding)

№	Options	*The dynamics of changes in the amount of N-NO_3 in the soil in different years, mg/kg*								
		2019 years			*2020 years*			*2021 years*		
		From planting seedlings before	*2- after feeding*	*4- after feeding*	*Before planting seedlings*	*2- after feeding*	*4- after feeding*	*Before planting seedlings*	*2- after feeding*	*4- after feeding*
1	$N_{350}P_{250}K_{170}$ control	75.2	56.3	37.8	74.4	44.2	35.2	76.5	67.0	45.8
2	$N_{350}P_{250}K_{170}$+Б10 (1)	96.1	80.2	42.4	96.7	87.0	53.0	99.7	91.5	56.1
3	$N_{350}P_{250}K_{170}$+Б20 (1+1)	93.6	109.0	37.5	96.6	106.1	45.0	102.0	108.7	60.7
4	$N_{350}P_{250}K_{170}$+Б30 (1+1+1)	94.1	116.2	61.8	98.0	115.7	80.0	100.6	131.5	89.4
5	$N_{350}P_{250}K_{170}$+Б40 (1+1+1+1)	95.7	119.4	74.4	96.2	122.0	84.2	98.0	135.0	98.6
6	$N_{350}P_{250}K_{170}$+Б50(1+1+1+1+1)	94.1	119.9	89.7	96.0	114.0	99.5	101.5	130.7	109.8
7	$N_{350}P_{250}K_{170}$+Б40(40)	99.0	82.7	42.3	98.1	80.0	55.1	97.0	111.3	63.7
8	Б50 (2+1+1+1)	112.0	158.4	148.1	114.1	175.1	149.3	116.4	194.1	147.6
9	Б80 (3+2+1+1+1)	149.8	179.1	99.2	146.0	163.5	106.1	138.0	177.1	102.3
10	$N_{350}P_{250}K_{170}$+H90 (3+2+2+2)	102.0	122.8	111.6	104.8	121.0	99.0	98.4	128.2	105.8

At the 2nd post-feeding stage of plant development, a significant increase in the amount of N-NOz in the soil of variants 3–6 fed with biohumus was observed. But during the three years of the study, while nitrogen in the soil was at the recommended level at the beginning of the experiment, we see a prominent decrease in all options as a result of its use for plant biomass from April. This situation is also related to the sharp increase in the release of nutrients from the crop by this time.

In addition, part of the nitrogen is washed away during irrigation and volatilizes due to denitrification. In this case, when only mineral fertilizer is used, even active feeding cannot maintain the optimal value of nitrogen. In all feeding options, where biohumus was used, the sharp reduction of nitrogen by the time of full ripening of the crop was avoided, and its amount was kept at stable optimal values during this period of plant development.

Unlike mineral fertilizer, nitrogen contained in biohumus is gradually mineralized and separated into the soil solution suitable for the plant's needs, so it can provide the plant with the same amount of nitrogen during the entire development period. As a result, the content of nitrate in the cucumber fruit is low and its quality is improved. In addition, nitrogen is not wasted due to leaching and denitrification during irrigation.

Throughout the experiment, in the control variant, the amount of nitrogen in the soil was maintained without reducing the recommended rate. However, a sharp decrease in nitrogen in the soil was observed at the stage of gross ripening of the crop. This is due to the consumption of a large amount of nitrogen during the harvest.

Therefore, at this stage, it is not possible to achieve the optimal value of nitrate in the soil by using only mineral fertilizers while maintaining a low content of nitrate in the fruit. In the options where biohumus was used, there was no sharp reduction of nitrogen even in the stage of gross yield of the cucumber plant, which was different from the option where only mineral fertilizer was used.

The analysis results from Table 8.14 reveal that, even when the quantity of biohumus added to the soil is multiplied by two or threefold (options 8–9), the amount of nitrogen in its content did not change dramatically. The smallest (20–21%) change in the amount of N-NOz in the time interval before the 2nd feeding was observed under the plow, in the 7th option, where the largest amount (4kg/m^2) of biohumus was used.

The patterns of changes in the amount of N-NOz in different options during the period of 60 days before the 2nd feeding remained almost unchanged in all years of the experiment. On the 60th day of the experiment (before the second feeding), the highest amount of N-NO_3 in the soil (66.3 and 68.4%) was observed in options 5–6. A sharp decrease in N-NO_3 content in the period before the second feeding was observed only in the soil of the control variant. A similar change was found in options 2–3, where one and two feedings with biohumus were carried out in the period up to the fourth feeding.

Thus, feeding with the same amount of biohumus several times in a row on the basis of the scheme allows to maintain the amount of N-NO_3 in the soil at optimal values.

Analyzing the dynamics of change in t N-NO_3 in the soil during the 150-day experiments showed that its amount decreases by 1.3–1.8 times at multiple stages involved in plant development. In the options where manure with mineral fertilizer was used, the superiority of N-NO_3 content over the control option remained at all stages of plant development. There was no significant difference in the change of N-NO_3 and N-NN_4 in the same options for 2019–2021.

The dynamics of changes in the amount of N-NN_4 in the composition of the greenhouse soil during the entire experiment are similar to the dynamics of changes in the amount of N-NO_3 under the same conditions, differing only in numerical values. Table 8.15 shows the dynamics of changes in the amount of N-NN_4 in the soil, which was tested according to different options.

Looking at the data in Table 8.15, it is evident that the most significant decline in ammonium content during various plant development stages occurred in the control group and in options 2 and 3, where early cessation of biohumus feeding was implemented.

On the other hand, the highest levels of N-NH_4 were sustained in variations 5–9, where continuous biohumus feeding was maintained. Therefore, providing multiple biohumus feedings during the growth period of cucumber plants contributes to the preservation of N-NO_3 and N-NH_4 levels in the soil, which is a critical factor in their growth.

From Table 8.16, we can see the patterns of changes in the amount of nitrate in the soil after harvesting in all years, as well as the changes in the amount of ammonium. In Tables 8.14 and 8.15, the lowest amount of N-NO_3 and N-NN_4 in the soil of the control variant corresponds to the third year of the experiment.

There is no significant difference in the patterns of changes in the amount of N-NO_3 and N-NN_4 in different years of the experiment. In all options, the amount of nitrate in the soil increases with the amount of biohumus used. Comparison of the results of using the same amount of biohumus (4kg/m^2) in two different ways in the table showed that dividing and using biohumus in the feeding process (option 5) is better than putting it under the plow at once (option 7).

When this method is used, we see that the amount of residual nitrate in the soil of variant 5 after the complete harvest is 3–3.5 times higher than the amount of residual nitrate in variant 7, where biohumus was used in the amount of 4 kg/m^2 under the plow (Table 8.16). In addition, as seen from the results of agrochemical analysis, there is no sharp change in the amount of NH_4+ and NO_3– in the soil during step feeding.

Thus, in the cultivation of cucumbers in greenhouse conditions, even distribution of nutrients and their effective use can be achieved by using

Table 8.15 Dynamics of Changes in the Amount of N-NN_4 in the Soil Fed with Biohumus of Different Rates (Post-feeding)

№	Options	*The dynamics of changes in the amount of N-NN_4 in the soil in different years, mg/kg*								
		2019 years			*2020 years*			*2021 years*		
		From planting seedlings before	*2- after feeding*	*4- after feeding*	*Before planting seedlings*	*2- after feeding*	*4- after feeding*	*Before planting seedlings*	*2- after feeding*	*4- after feeding*
1	$N_{350}P_{250}K_{170}$ control	21.3	7.6	5.5	30.1	7.8	4.7	18.0	15.6	6.7
2	$N_{350}P_{250}K_{170}$+Б10 (1)	18.0	15.6	4.8	23.4	15.3	9.1	28.4	26.0	13.9
3	$N_{350}P_{250}K_{170}$+Б20 (1+1)	26.9	19.0	8.7	26.9	29.3	8.9	28.5	36.9	15.5
4	$N_{350}P_{250}K_{170}$+Б30 (1+1+1)	27.0	18.3	8.9	27.0	28.3	9.1	28.5	38.6	10.6
5	$N_{350}P_{250}K_{170}$+Б40 (1+1+1+1)	24.5	31.3	17.4	27.5	37.4	19.4	28.2	42.4	29.0
6	$N_{350}P_{250}K_{170}$+Б50(1+1+1+1+1)	27.0	26.8	21.3	28.1	37.3	20.3	27.6	37.2	30.4
7	$N_{350}P_{250}K_{170}$+Б40(40)	23.2	24.7	13.8	28.1	34.1	13.6	36.2	33.5	12.4
8	Б50 (2+1+1+1)	38.3	63.4	6.9	42.5	53.0	36.7	46.2	53.9	48.0
9	Б80 (3+2+1+1+1)	44.6	26.8	16.8	51.0	46.6	17.6	66.0	47.0	37.7
10	$N_{350}P_{250}K_{170}$+Н90 (3+2+2+2)	28.1	41.2	22.4	30.3	21.0	20.0	26.1	31.0	22.2

Table 8.16 The Evolution of Nm, N-NO_3, and N-NH_4 Quantities in Greenhouse Soil Following the Complete Harvest in the Years 2019–2021

№	Options	*The dynamics of changes in the amount of Nm, N-NO_3 and N-NN_4 in the soil in different years at the end of the growing season,* мг/кг											
		Nm. change in quantity				*Change in amount of N-NO_3*				*Change in amount of N-NN_4*			
		2019 years	*2020 years*	*2021 years*	*2019–2021 year*	*2019 years*	*2020 years*	*2021 years*	*2019–2021 years*	*2019 years*	*2020 years*	*2021 years*	*2019–2021 years*
1	$N_{350}P_{250}K_{170}$	24.9	19.6	14.8	19.7	19.3	15.2	11.5	15.3	5.6	4.4	3.3	4.4
2	$N_{350}P_{250}K_{170}$+Б10 (1)	22.8	18.9	18.2	19.9	17.7	14.7	14.1	15.5	5.1	4.2	4.1	4.5
3	$N_{350}P_{250}K_{170}$+Б20 (1+1)	27.8	29.5	33.9	30.4	21.6	22.9	26.3	23.6	6.2	6.6	7.6	6.8
4	$N_{350}P_{250}K_{170}$+Б30 (1+1+1)	45.7	49.4	53.8	49.6	35.5	38.4	41.8	38.5	10.2	11.0	12.0	11.1
5	$N_{350}P_{250}K_{170}$+Б40 (1+1+1+1)	70.9	76.6	81.0	76.2	55.1	59.5	62.9	59.2	15.8	17.1	18.1	17.0
6	$N_{350}P_{250}K_{170}$+Б50(1+1+1+1+1)	131.3	137.8	145.4	137.8	102	107	113	107	29.3	30.8	32.4	30.8
7	$N_{350}P_{250}K_{170}$+Б40(40)	25.5	27.2	19.6	24.1	19.8	21.1	15.2	18.7	5.7	6.1	4.4	5.4
8	Б50 (2+1+1+1)	24.0	29.7	23.1	25.6	18.6	23.1	17.9	19.9	5.4	6.6	5.2	5.7
9	Б80 (3+2+1+1+1)	25.5	17.2	19.6	20.8	19.8	13.4	15.2	16.2	5.7	3.8	4.4	4.6
10	$N_{350}P_{250}K_{170}$+Н90 (3+2+2+2)	50.8	52.5	54.9	52.7	39.5	40.8	42.7	40.9	11.3	11.7	12.2	11.8

biohumus. For the best results, using 5 kg of biohumus per 1 m^2 of the area from planting to the end of the yield period 5 times at the rate of 1 kg/m^2 gives optimal results in terms of nutrients in the soil. When repeated feeding with biohumus is carried out due to the slow release of nitrogen from organic fertilizer, a normal regime of nitrogen nutrition of the plant is provided compared to the use of mineral fertilizers, and the soil with added biohumus fully meets all the requirements for growing cucumbers in greenhouse conditions (Plekhanova and Savelyeva, 1997; Voitovich et al., 2002).

8.3.3 The effect of biohumus on heavy metals in soil in greenhouse conditions

The effect of biohumus on the amount of heavy metals in the soil of the greenhouse where cucumbers were grown was controlled during the three-year experiments. Currently, the maximum permissible limit (REM) of most metals in soil and agricultural products has been approved.

The amount of heavy metals in agricultural products, including cucumbers, under greenhouse conditions depends more on their amount in mobile form than on their gross amount in the soil. Hence, the primary approach for evaluating the degree of heavy metal contamination in greenhouse soil relies on the measurement of their mobile form, as established by Plekhanova et al. (1995b). Several soil categorizations based on the content of mobile heavy metals exist, and the classifications developed by Obukhov and Renks are provided in Tables 8.17–8.18.

In this duration of 3 years of research, after harvesting and completing the experiment, the amount of Cd, Co, Cr, Cu, Mn, Ni, Pb, Zn and Mo in the plow layer of the soil of the experimental site in mobile form (mg/kg), biohumus added to the soil (kg/m^2)) effects were examined. In the experiment, methods widely used in the practice of determining heavy metals were used.

Table 8.17 Classification of Soil by the Amount and Pollution of Heavy Metals in Mobile form (Compared to Air-Dry Soil), mg/kg

Gradations	*Pb*	*Pb*	*Cd*	*Cu*	*Ni*	*Co*
Amount						
Жуда кам	<0.2	<0.02	<1	<0.2	<0.2	<0.1
Кам	0.2–0.5	0.02–0.05	1–2	0.2–0.5	0.2–0.5	0.1–0.2
Ўртача таминланган	0.5–1.5	0.05–0.10	2–5	0.5–2.0	0.5–1.5	0.2–0.5
Юқори	1.5–5.0	0.10–0.50	5–20	2.0–5.0	1.5–5.0	0.5–3.0
Pollution						
Кучсиз (РЭМ)	5–10	0.5–1.0	20–50	5–10	5–10	3–5
Ўртача	10–50	1.0–3.0	50–100	10–50	10–50	5–25
Кучли	50–100	3.0–5.0	100–200	50–100	50–100	25–50
Жуда кучли	>100	>5.0	>200	>100	>100	>50

Source: Obukhov and Popova (1992)

Table 8.18 The Amount of Trace Elements in the Soil According to Rinkis.

	Amount of elements in soil, mg/kg					
R	*B*	*Cu*	*Mn*	*Mo*	*Zn*	*Co*
Very little	<0.1	<0.3	<1	<0.05	<0.2	<0.2
Few	0.1–0.2	0.3–1.5	1–10	0.05–0.15	0.2–1.0	0.2–1.0
Average Provided	0.3–0.5	2–3	20–50	0.20–0.25	2–3	1.5–3.0
High	0.6–1	4–7	60–100	0.3–0.5	4–5	4–5
Very high	>1	>7	>100	>0.5	>5	>5

Table 8.19 Findings from the Analysis of Heavy Metal Levels in Soil after Harvest (Averaged over the Years 2019–2021)

		Amount of heavy metals in mobile form in harvested soil, mg/kg								
№	*Biohumus standard, kg/m²*	*Cd*	*Co*	*Cr*	*Cu*	*Mn*	*Ni*	*Pb*	*Zn*	Mo
1	$N_{350}P_{250}K_{170}$	0.38	0.45	1.53	2.9	46.6	0.84	4.89	44.37	0.16
2	$N_{350}P_{250}K_{170}$+$Б_{10}$(1)	0.31	0.25	1.40	2.48	41.90	0.76	3.25	32.78	0.16
3	$N_{350}P_{250}K_{170}$+$Б_{20}$(1+1)	0.18	0.19	1.29	2.30	36.32	0.77	2.23	22.1	0.14
4	$N_{350}P_{250}K_{170}$+$Б_{30}$(1+1+1)	0.10	0.13	1.12	1.09	28.54	0.59	1.26	13.16	0.12
5	$N_{350}P_{250}K_{170}$+$Б_{40}$(1+1+1+1)	0.08	0.15	1.06	1.25	28.69	0.65	1.17	12.91	0.11
6	$N_{350}P_{250}K_{170}$+$Б_{50}$(1+1+1+1+1)	0.08	0.12	1.06	1.03	28.63	0.58	1.35	12.4	0.11
7	$Б_{50}$(2+1+1+1)	0.07	0.08	0.80	0.87	22.53	0.57	1.13	9.93	0.10
8	$Б_{80}$(3+2+1+1+1)	0.07	0.10	0.84	0.68	22.31	0.54	1.02	9.45	0.09
9	$N_{350}P_{250}K_{170}$+$Б_{40}$(40)	0.06	0.07	0.59	0.66	18.03	0.36	0.71	8.01	0.07
10	$N_{350}P_{250}K_{170}$+H_{90}(3+2+2+2)	0.33	0.34	1.44	2.46	41.48	0.72	3.49	37.17	0.18

Table 8.19 shows the three-year average value of heavy metals in the soil of the greenhouse after the harvest of 2019–2021.

In all the cases where the experiment was conducted, the amount of heavy metals detected in the soil of the control option is lower than their permissible norm in the soil approved by hygienic regulations (Higienicheskie normativity GN 2.1.7.2042–06 "Orientirovochno dopustimye koncentratsii khimicheskix veshchestv v pochve"). The amount of heavy metals in the soil of the control variant without biohumus in the three-year experiments: Cd-0.34–0.42 mg/kg; Co–0.39–0.50 mg/kg; Cr–1.48–1.59 mg/kg; Cu–2.82–2.91 mg/kg; Mn-45.80–47.2 mg/kg; Ni–0.81–0.87 mg/kg; Pb-4.75–5.07 mg/kg; Zn–41.97–46.00 mg/kg; Mo was 0.17–0.14 mg/kg (Appendix 8.11).

Therefore, according to the amount of heavy metals, the soil of the experimental area without adding biohumus belongs to the category of lightly contaminated soil with lead and spirit, highly supplied with cadmium, chromium, copper and moderately supplied with cobalt, manganese, nickel and molybdenum (Table 8.18).

From the results presented in Table 8.19, we can see that under the influence of biohumus applied to the soil, the content of heavy metals in mobile form decreases. From the analysis of the mentioned three-year studies (Table 8.19), it can be concluded that the effect of biohumus on heavy metals in the soil in greenhouse conditions is not the same for all metals but is specific for each metal.

A sharp decrease in the amount of some metals was observed under the effect of the same amount of added biohumus, while the decrease in other metals was relatively low. In this case, the general situation for all metals is that the amount of metal ions in the mobile form decreases with an inverse relation to amount of biohumus that was applied to soil.

It was noted that metal lead decreased the most in accordance with the increase in biohumus rate. Its three-year average amount in the control option without biohumus is equal to 4.89 mg/kg, and this soil belongs to the category of soil lightly polluted with lead according to the amount of heavy metals and pollution classification. During the three-year experiments, it was observed that the amount of lead in the soil was reduced by 1.5–7.1 times under the influence of 1–5 kg/m^2 of biohumus (Appendix 8.11). As a result, the minimum amount of lead in the soil after the use of biohumus was 0.71mg/kg. According to Tables 8.17 and 8.18, it can be seen that such soil belongs to the category of soil with moderate lead supply.

From the experimental results presented in Table 8.19, we can see that the amount of cadmium, cobalt and zinc in the soil is significantly reduced under the influence of biohumus. The soil of the control option, where only mineral fertilizer was applied, belongs to the soil category of weakly contaminated or highly enriched with heavy metals according to the content of cadmium, cobalt and zinc in it.

During the experiment, under the influence of biohumus applied up to 5 kg/m^2, the amount of the above-mentioned metals decreased from 5.5 to 6.6 times, and it was noted that the soil of the experimental area belongs to the category of soil with moderate or low supply of cadmium, cobalt and zinc. It was found that under the influence of biohumus with a rate of up to 5 kg/m^2, the amount of copper in mobile form in the greenhouse soil decreased by 4.5 times compared to its amount in the control option.

Molybdenum, manganese, and chromium were the least reduced metals under the influence of applied biohumus at the same rate, their decreased part under the influence of 5 kg/m^2 biohumus was 2.3–2.6 times of the initial amount.

After applying 3–5 kg/m^2 of biohumus, the concentration of transportable heavy metals in the soil was found to have decreased, resulting in the soil reaching a moderate level of metal availability. Consequently, biohumus serves as a sorbent for the cations of heavy metals present in the soil and improves the soil's ability to act as a buffer. The results observed in the studies confirm that it is possible to improve the quality of soil contaminated with heavy metals in closed soil conditions by adding biohumus to it by attaching heavy metals to the organic phase.

The incorporation of biohumus in cucumber cultivation significantly decreases the concentration of lead, cadmium, cobalt, zinc, and copper in the readily available form within the greenhouse soil.

Appendices

Appendix 8.1 Agrochemical Examination of Greenhouse Soil Conducted Before Planting, Treated With Both Mineral and Organic Fertilizers, Which Include Biohumus and Gunk

№	Options	pH	Amount of NPK in the soil before transplanting, mg/kg		
			Nм	P_2O_5	K_2O
2019 years					
1	$N_{350}P_{250}K_{170}$	6.8	96.5	58.5	323.5
2	$N_{350}P_{250}K_{170}$+Б$_{10}$(1)	6.7	114.1	70.7	331.8
3	$N_{350}P_{250}K_{170}$+Б$_{20}$(1+1)	6.9	120.5	71.1	327.0
4	$N_{350}P_{250}K_{170}$+Б$_{30}$(1+1+1)	6.8	121.1	73.0	322.8
5	$N_{350}P_{250}K_{170}$+Б$_{40}$(1+1+1+1)	6.9	120.2	71.7	319.9
6	$N_{350}P_{250}K_{170}$+Б$_{50}$) (1+1+1+1+1)	6.9	121.1	70.7	327.8
7	Б$_{50}$(2+1+1+1)	7.0	122.2	70.4	339.5
8	Б$_{80}$(3+2+1+1+1)	7.1	150.3	80.1	364.0
9	$N_{350}P_{250}K_{170}$+Б$_{40}$(40)	7.0	194.4	108.0	422.0
10	$N_{350}P_{250}K_{170}+H_{90}$(3+2+2+2)	6.6	130.1	70.6	340.1
2020 years					
1	$N_{350}P_{250}K_{170}$	6.7	104.5	60.0	322.5
2	$N_{350}P_{250}K_{170}$+Б$_{10}$(1)	6.8	120.1	73.9	344.8
3	$N_{350}P_{250}K_{170}$+Б$_{20}$(1+1)	6.9	123.5	76.1	337.0
4	$N_{350}P_{250}K_{170}$+Б$_{30}$(1+1+1)	6.9	125.1	78.0	332.8
5	$N_{350}P_{250}K_{170}$+Б$_{40}$(1+1+1+1)	6.9	123.2	75.7	329.9
6	$N_{350}P_{250}K_{170}$+Б$_{50}$) (1+1+1+1+1)	6.9	124.1	72.7	337.8
7	Б$_{50}$(2+1+1+1)	7.0	126.2	77.4	348.5
8	Б$_{80}$(3+2+1+1+1)	7.1	156.6	83.1	378.1
9	$N_{350}P_{250}K_{170}$+Б$_{40}$(40)	7.2	197.0	119.8	432.0
10	$N_{350}P_{250}K_{170}+H_{90}$(3+2+2+2)	6.7	135.1	68.6	334.1
2021 years					
1	$N_{350}P_{250}K_{170}$	6.5	94.5	63.7	318.5
2	$N_{350}P_{250}K_{170}$+Б$_{10}$(1)	6.7	128.1	78.7	356.1

Appendix 8.1 (Continued) Agrochemical Examination of Greenhouse Soil Conducted Before Planting, Treated With Both Mineral and Organic Fertilizers, Which Include Biohumus and Gunk

№	*Options*	*pH*	*Amount of NPK in the soil before transplanting, mg/kg*		
			Nм	P_2O_5	K_2O
3	$N_{350}P_{250}K_{170}$+$Б_{20}$(1+1)	6.9	130.5	78.1	341.0
4	$N_{350}P_{250}K_{170}$+$Б_{30}$(1+1+1)	6.8	129.1	89.0	341.8
5	$N_{350}P_{250}K_{170}$+$Б_{40}$(1+1+1+1)	6.9	126.2	77.7	340.9
6	$N_{350}P_{250}K_{170}$+$Б_{50}$) (1+1+1+1+1)	6.9	129.1	75.7	344.8
7	$Б_{50}$(2+1+1+1)	7.1	133.2	86.4	350.5
8	$Б_{80}$(3+2+1+1+1)	7.2	162.6	88.1	386.0
9	$N_{350}P_{250}K_{170}$+$Б_{40}$(40)	7.1	204.0	121.8	438.0
10	$N_{350}P_{250}K_{170}$+H_{90}(3+2+2+2)	6.6	124.5	65.7	328.5

Appendix 8.2 The Result of Soil Analysis Before Picking the First Crop (NPK Amount, mg/kg)

№	*Options*	*pH*	*NPK available in greenhouse soil before first feeding, mg/kg*		
			N	P_2O_5	K_2O
2019 years					
1	$N_{350}P_{250}K_{170}$	6.6	72.4	43.4	198.0
2	$N_{350}P_{250}K_{170}$+$Б_{10}$(1)	6.9	87.1	52.1	226.8
3	$N_{350}P_{250}K_{170}$+$Б_{20}$(1+1)	6.9	84.5	55.7	219.5
4	$N_{350}P_{250}K_{170}$+$Б_{30}$(1+1+1)	6.7	85.1	52.7	209.7
5	$N_{350}P_{250}K_{170}$+$Б_{40}$(1+1+1+1)	6.9	82.3	55.1	222.8
6	$N_{350}P_{250}K_{170}$+$Б_{50}$) (1+1+1+1+1)	6.9	91.1	51.9	212.3
7	$Б_{50}$(2+1+1+1)	6.9	83.2	50.4	229.5
8	$Б_{80}$(3+2+1+1+1)	7.1	108.3	61.1	249.0
9	$N_{350}P_{250}K_{170}$+$Б_{40}$(40)	7.0	142.4	95.8	227.3
10	$N_{350}P_{250}K_{170}$+H_{90}(3+2+2+2)	6.7	111.2	52.6	225.1
2020 years					
1	$N_{350}P_{250}K_{170}$	6.7	73.2	46.6	209.6
2	$N_{350}P_{250}K_{170}$+$Б_{10}$(1)	6.7	92.9	56.9	230.6
3	$N_{350}P_{250}K_{170}$+$Б_{20}$(1+1)	6.8	86.5	58.6	232.1
4	$N_{350}P_{250}K_{170}$+$Б_{30}$(1+1+1)	6.9	89.6	57.0	222.8
5	$N_{350}P_{250}K_{170}$+$Б_{40}$(1+1+1+1)	6.9	86.2	59.3	230.4
6	$N_{350}P_{250}K_{170}$+$Б_{50}$) (1+1+1+1+1)	6.9	94.8	55.2	235.1
7	$Б_{50}$(2+1+1+1)	7.0	88.3	54.2	236.5

(Continued)

Appendix 8.2 (Continued) The Result of Soil Analysis Before Picking the First Crop (NPK Amount, mg/kg)

№	*Options*	*pH*	*NPK available in greenhouse soil before first feeding, mg/kg*		
			N	*P_2O_5*	*K_2O*
8	$Б_{80}(3+2+1+1+1)$	7.1	117.0	64.0	259.3
9	$N_{350}P_{250}K_{170}+Б_{40}(40)$	7.0	148.0	102.2	290.8
10	$N_{350}P_{250}K_{170}+H_{90}(3+2+2+2)$	6.7	94.6	56.8	217.2
2021 years					
1	$N_{350}P_{250}K_{170}$	6.8	70.9	41.0	223.0
2	$N_{350}P_{250}K_{170}+Б_{10}(1)$	6.7	97.3	59.0	242.3
3	$N_{350}P_{250}K_{170}+Б_{20}(1+1)$	6.9	93.4	65.1	238.7
4	$N_{350}P_{250}K_{170}+Б_{30}(1+1+1)$	6.9	96.8	59.2	242.3
5	$N_{350}P_{250}K_{170}+Б_{40}(1+1+1+1)$	6.9	92.4	67.2	238.6
6	$N_{350}P_{250}K_{170}+Б_{50})$ (1+1+1+1+1)	6.9	99.3	58.4	250.9
7	$Б_{50}(2+1+1+1)$	7.1	96.9	58.1	263.4
8	$Б_{80}(3+2+1+1+1)$	7.1	124.0	65.7	273.2
9	$N_{350}P_{250}K_{170}+Б_{40}(40)$	7.0	156.0	107.4	306.6
10	$N_{350}P_{250}K_{170}+H_{90}(3+2+2+2)$	6.8	100.9	54.0	213.0

Appendix 8.3 The Quantity of Nutrients That Have Accrued in the Greenhouse Soil Following the Initial Fertilization

№	*Options*	*pH*	*Amount of NPK in greenhouse soil after the first feeding, mg/kg*		
			Nм	*P_2O_5*	*K_2O*
2019 years					
1	$N_{350}P_{250}K_{170}$	6.8	101.4	62.4	221.0
2	$N_{350}P_{250}K_{170}+Б_{10}(1)$	6.9	131.1	81.1	270.7
3	$N_{350}P_{250}K_{170}+Б_{20}(1+1)$	6.9	151.2	94.8	288.5
4	$N_{350}P_{250}K_{170}+Б_{30}(1+1+1)$	7.0	150.2	92.0	291.5
5	$N_{350}P_{250}K_{170}+Б_{40}(1+1+1+1)$	6.9	161.2	92.8	296.5
6	$N_{350}P_{250}K_{170}+Б_{50})$ (1+1+1+1+1)	6.9	158.2	102.8	286.5
7	$Б_{50}(2+1+1+1)$	7.2	148.4	82.9	301.0
8	$Б_{80}(3+2+1+1+1)$	7.1	203.7	120.8	361.1
9	$N_{350}P_{250}K_{170}+Б_{40}(40)$	7.2	241.6	152.5	383.5
10	$N_{350}P_{250}K_{170}+H_{90}(3+2+2+2)$	6.5	183.4	87.3	302.5

Appendix 8.3 (Continued) The Quantity of Nutrients That Have Accrued in the Greenhouse Soil Following the Initial Fertilization

№	*Options*	*pH*	*Amount of NPK in greenhouse soil after the first feeding, mg/kg*		
			Nм	*P_2O_5*	*K_2O*
2020 years					
1	$N_{350}P_{250}K_{170}$	6.8	102.2	65.6	232.6
2	$N_{350}P_{250}K_{170}$+$Б_{10}$(1)	6.7	136.9	88.9	280.5
3	$N_{350}P_{250}K_{170}$+$Б_{20}$(1+1)	6.8	159.2	96.7	295.1
4	$N_{350}P_{250}K_{170}$+$Б_{30}$(1+1+1)	6.9	155.7	99.3	289.6
5	$N_{350}P_{250}K_{170}$+$Б_{40}$(1+1+1+1)	6.9	166.1	100.0	293.1
6	$N_{350}P_{250}K_{170}$+$Б_{50}$) (1+1+1+1+1)	6.9	161.9	109.1	297.3
7	$Б_{50}$(2+1+1+1)	7.0	153.5	88.7	313.0
8	$Б_{80}$(3+2+1+1+1)	7.1	212.4	123.7	378.4
9	$N_{350}P_{250}K_{170}$+$Б_{40}$(40)	7.0	237.2	148.9	397.0
10	$N_{350}P_{250}K_{170}$+H_{90}(3+2+2+2)	6.5	166.8	84.5	284.6
2021 years					
1	$N_{350}P_{250}K_{170}$	6.8	99.9	70.0	246.0
2	$N_{350}P_{250}K_{170}$+$Б_{10}$(1)	6.8	146.3	91.0	292.2
3	$N_{350}P_{250}K_{170}$+$Б_{20}$(1+1)	6.9	158.1	103.2	304.7
4	$N_{350}P_{250}K_{170}$+$Б_{30}$(1+1+1)	7.0	168.9	101.5	302.1
5	$N_{350}P_{250}K_{170}$+$Б_{40}$(1+1+1+1)	6.9	170.3	108.9	297.3
6	$N_{350}P_{250}K_{170}$+$Б_{50}$) (1+1+1+1+1)	6.9	166.4	118.3	315.1
7	$Б_{50}$(2+1+1+1)	7.1	162.1	89.6	321.9
8	$Б_{80}$(3+2+1+1+1)	7.1	229.4	130.4	395.3
9	$N_{350}P_{250}K_{170}$+$Б_{40}$(40)	7.0	245.2	154.1	412.8
10	$N_{350}P_{250}K_{170}$+H_{90}(3+2+2+2)	6.6	143.1	82.7	290.4

Appendix 8.4 The NPK Concentration Within the Greenhouse Soil Prior to the Second Plant Feeding, Occurring During the Phase of High Productivity

№	*Options*	*pH*	*Amount of NPK present in greenhouse soil before second feeding, mg/kg*		
			Nм	*P_2O_5*	*K_2O*
2019 years					
1	$N_{350}P_{250}K_{170}$	6.9	35.7	25.6	113.6
2	$N_{350}P_{250}K_{170}$+$Б_{10}$(1)	6.7	49.3	41.9	123.8
3	$N_{350}P_{250}K_{170}$+$Б_{20}$(1+1)	6.9	64.4	49.2	140.5

(*Continued*)

Appendix 8.4 (Continued) The NPK Concentration Within the Greenhouse Soil Prior to the Second Plant Feeding, Occurring During the Phase of High Productivity

№	*Options*	*pH*	*Amount of NPK present in greenhouse soil before second feeding, mg/kg*		
			Nм	P_2O_5	K_2O
4	$N_{350}P_{250}K_{170}$+$Б_{30}$(1+1+1)	6.8	68.3	50.1	148.2
5	$N_{350}P_{250}K_{170}$+$Б_{40}$(1+1+1+1)	6.9	76.4	45.4	143.6
6	$N_{350}P_{250}K_{170}$+$Б_{50}$) (1+1+1+1+1)	6.9	71.4	54.3	150.2
7	$Б_{50}$(2+1+1+1)	7.0	65.5	38.0	149.0
8	$Б_{80}$(3+2+1+1+1)	7.1	105.0	66.4	230.8
9	$N_{350}P_{250}K_{170}$+$Б_{40}$(40)	7.0	153.0	98.0	262.9
10	$N_{350}P_{250}K_{170}$+H_{90}(3+2+2+2)	6.6	91.1	36.0	132.3
2020 years					
1	$N_{350}P_{250}K_{170}$	6.7	32.9	21.3	103.1
2	$N_{350}P_{250}K_{170}$+$Б_{10}$(1)	6.6	58.8	47.6	150.6
3	$N_{350}P_{250}K_{170}$+$Б_{20}$(1+1)	6.7	73.9	52.2	146.7
4	$N_{350}P_{250}K_{170}$+$Б_{30}$(1+1+1)	6.8	77.4	54.6	159.0
5	$N_{350}P_{250}K_{170}$+$Б_{40}$(1+1+1+1)	6.9	72.8	55.7	158.7
6	$N_{350}P_{250}K_{170}$+$Б_{50}$) (1+1+1+1+1)	6.9	74.7	64.7	168.8
7	$Б_{50}$(2+1+1+1)	7.0	68.8	40.6	152.6
8	$Б_{80}$(3+2+1+1+1)	7.0	121.1	70.9	237.2
9	$N_{350}P_{250}K_{170}$+$Б_{40}$(40)	6.9	161.0	104.1	217.7
10	$N_{350}P_{250}K_{170}$+H_{90}(3+2+2+2)	6.7	74.0	27.5	127.4
2021 years					
1	$N_{350}P_{250}K_{170}$	6.8	37.6	50.6	123.6
2	$N_{350}P_{250}K_{170}$+$Б_{10}$(1)	6.7	71.0	57.4	182.3
3	$N_{350}P_{250}K_{170}$+$Б_{20}$(1+1)	6.9	94.1	65.2	194.7
4	$N_{350}P_{250}K_{170}$+$Б_{30}$(1+1+1)	6.8	93.5	67.8	188.5
5	$N_{350}P_{250}K_{170}$+$Б_{40}$(1+1+1+1)	6.9	100.8	70.1	201.6
6	$N_{350}P_{250}K_{170}$+$Б_{50}$) (1+1+1+1+1)	6.9	91.3	79.5	204.4
7	$Б_{50}$(2+1+1+1)	7.0	79.5	49.3	199.8
8	$Б_{80}$(3+2+1+1+1)	7.1	139.0	89.4	244.6
9	$N_{350}P_{250}K_{170}$+$Б_{40}$(40)	7.0	175.0	110.0	209.3
10	$N_{350}P_{250}K_{170}$+H_{90}(3+2+2+2)	6.7	81.2	44.7	167.6

Appendix 8.5 The Total Nutrient Content That Has Amassed in the Greenhouse Soil Following the Second Feeding

№	*Options*	*pH*	*Amount of total NPK accumulated after the second feeding, mg/kg*		
			Nм	*P_2O_5*	*K_2O*
2019 years					
1	$N_{350}P_{250}K_{170}$	6.8	63.9	44.6	146.8
2	$N_{350}P_{250}K_{170}$+$Б_{10}$(1)	6.9	95.8	62.1	177.3
3	$N_{350}P_{250}K_{170}$+$Б_{20}$(1+1)	6.9	128.0	75.9	216.8
4	$N_{350}P_{250}K_{170}$+$Б_{30}$(1+1+1)	6.8	134.5	103.0	245.1
5	$N_{350}P_{250}K_{170}$+$Б_{40}$(1+1+1+1)	6.9	150.7	93.0	248.1
6	$N_{350}P_{250}K_{170}$+$Б_{50}$) (1+1+1+1+1)	7.0	146.7	102.0	245.0
7	$Б_{50}$(2+1+1+1)	7.0	107.4	63.3	228.8
8	$Б_{80}$(3+2+1+1+1)	7.1	221.8	118.0	366.0
9	$N_{350}P_{250}K_{170}$+$Б_{40}$(40)	7.0	205.9	142.0	315.7
10	$N_{350}P_{250}K_{170}$+H_{90}(3+2+2+2)	6.5	164.0	72.5	235.7
2020 years					
1	$N_{350}P_{250}K_{170}$	6.8	51.9	40.7	126.4
2	$N_{350}P_{250}K_{170}$+$Б_{10}$(1)	6.7	102.3	71.7	194.3
3	$N_{350}P_{250}K_{170}$+$Б_{20}$(1+1)	6.9	135.4	91.7	221.3
4	$N_{350}P_{250}K_{170}$+$Б_{30}$(1+1+1)	6.8	144.0	108.8	258.4
5	$N_{350}P_{250}K_{170}$+$Б_{40}$(1+1+1+1)	6.9	159.4	104.9	254.1
6	$N_{350}P_{250}K_{170}$+$Б_{50}$) (1+1+1+1+1)	6.9	151.3	108.9	268.2
7	$Б_{50}$(2+1+1+1)	7.0	114.1	76.1	239.1
8	$Б_{80}$(3+2+1+1+1)	7.1	228.1	125.2	381.2
9	$N_{350}P_{250}K_{170}$+$Б_{40}$(40)	7.0	210.1	132.3	322.8
10	$N_{350}P_{250}K_{170}$+H_{90}(3+2+2+2)	6.7	142.0	61.4	191.7
2021 years					
1	$N_{350}P_{250}K_{170}$	6.6	82.6	70.0	176.9
2	$N_{350}P_{250}K_{170}$+$Б_{10}$(1)	6.9	117.5	91.5	216.0
3	$N_{350}P_{250}K_{170}$+$Б_{20}$(1+1)	6.9	145.6	100.7	249.3
4	$N_{350}P_{250}K_{170}$+$Б_{30}$(1+1+1)	6.8	170.1	117.0	277.9
5	$N_{350}P_{250}K_{170}$+$Б_{40}$(1+1+1+1)	7.1	177.4	119.3	271.0
6	$N_{350}P_{250}K_{170}$+$Б_{50}$) (1+1+1+1+1)	6.9	167.9	128.7	283.8
7	$Б_{50}$(2+1+1+1)	7.0	144.8	94.8	276.3
8	$Б_{80}$(3+2+1+1+1)	7.1	248.0	143.7	408.6
9	$N_{350}P_{250}K_{170}$+$Б_{40}$(40)	6.9	224.1	148.2	344.4
10	$N_{350}P_{250}K_{170}$+H_{90}(3+2+2+2)	6.6	159.2	78.6	261.9

Appendix 8.6 The NPK Content in the Greenhouse Soil Prior to the Third Feeding, Which Occurred During the Phase of High Plant Productivity

№	*Options*	*pH*	*Amount of NPK present in the greenhouse soil before the third feeding, mg/kg*		
			Nм	*P_2O_5*	*K_2O*
2019 years					
1	$N_{350}P_{250}K_{170}$	6.8	19.7	20.9	76.9
2	$N_{350}P_{250}K_{170}$+$Б_{10}$(1)	6.9	36.5	36.6	102.7
3	$N_{350}P_{250}K_{170}$+$Б_{20}$(1+1)	6.9	33.0	41.8	107.1
4	$N_{350}P_{250}K_{170}$+$Б_{30}$(1+1+1)	6.8	55.5	51.2	93.2
5	$N_{350}P_{250}K_{170}$+$Б_{40}$(1+1+1+1)	6.9	50.8	50.7	92.2
6	$N_{350}P_{250}K_{170}$+$Б_{50}$) (1+1+1+1+1)	6.8	52.7	61.9	95.3
7	$Б_{50}$(2+1+1+1)	7.0	32.9	32.5	93.9
8	$Б_{80}$(3+2+1+1+1)	7.1	135.0	86.0	216.5
9	$N_{350}P_{250}K_{170}$+$Б_{40}$(40)	7.0	100.0	83.7	150.6
10	$N_{350}P_{250}K_{170}$+H_{90}(3+2+2+2)	6.6	90.7	22.5	102.3
2020 years					
1	$N_{350}P_{250}K_{170}$	6.8	16.6	24.1	44.0
2	$N_{350}P_{250}K_{170}$+$Б_{10}$(1)	6.7	38.2	44.4	111.9
3	$N_{350}P_{250}K_{170}$+$Б_{20}$(1+1)	6.8	40.1	45.5	118.6
4	$N_{350}P_{250}K_{170}$+$Б_{30}$(1+1+1)	6.8	72.3	58.2	103.6
5	$N_{350}P_{250}K_{170}$+$Б_{40}$(1+1+1+1)	7.0	66.6	58.5	105.8
6	$N_{350}P_{250}K_{170}$+$Б_{50}$) (1+1+1+1+1)	6.9	61.9	70.7	120.3
7	$Б_{50}$(2+1+1+1)	7.0	38.9	37.6	105.9
8	$Б_{80}$(3+2+1+1+1)	7.1	139.0	93.6	225.0
9	$N_{350}P_{250}K_{170}$+$Б_{40}$(40)	7.0	109.9	96.1	160.0
10	$N_{350}P_{250}K_{170}$+H_{90}(3+2+2+2)	6.7	77.5	29.8	72.9
2021 years					
1	$N_{350}P_{250}K_{170}$	6.8	22.8	30.1	70.1
2	$N_{350}P_{250}K_{170}$+$Б_{10}$(1)	6.9	49.6	53.8	135.2
3	$N_{350}P_{250}K_{170}$+$Б_{20}$(1+1)	6.9	46.5	56.1	134.3
4	$N_{350}P_{250}K_{170}$+$Б_{30}$(1+1+1)	6.8	81.8	64.8	124.4
5	$N_{350}P_{250}K_{170}$+$Б_{40}$(1+1+1+1)	6.9	86.5	65.8	112.9
6	$N_{350}P_{250}K_{170}$+$Б_{50}$) (1+1+1+1+1)	7.0	78.8	76.1	128.8
7	$Б_{50}$(2+1+1+1)	6.9	47.5	43.0	114.4
8	$Б_{80}$(3+2+1+1+1)	7.1	146.0	117.0	258.1
9	$N_{350}P_{250}K_{170}$+$Б_{40}$(40)	7.0	126.0	111.0	193.9
10	$N_{350}P_{250}K_{170}$+H_{90}(3+2+2+2)	6.5	79.0	50.2	110.1

Appendix 8.7 The Total NPK Content Amassed in the Greenhouse Soil Following the Third Plant Feeding, Occurring During the Phase of High Productivity

№	*Options*	*pH*	*The amount of NPK accumulated in the soil after the third feeding, mg/kg*		
			Nм	*P_2O_5*	*K_2O*
2019 years					
1	$N_{350}P_{250}K_{170}$	6.8	46.7	40.3	90.2
2	$N_{350}P_{250}K_{170}$+Б$_{10}$(1)	6.8	68.0	57.6	99.5
3	$N_{350}P_{250}K_{170}$+Б$_{20}$(1+1)	6.8	67.0	57.5	114.3
4	$N_{350}P_{250}K_{170}$+Б$_{30}$(1+1+1)	6.9	99.5	86.3	161.3
5	$N_{350}P_{250}K_{170}$+Б$_{40}$(1+1+1+1)	6.9	129.9	91.5	165.2
6	$N_{350}P_{250}K_{170}$+Б$_{50}$) (1+1+1+1+1)	6.9	131.8	92.7	188.3
7	Б$_{50}$(2+1+1+1)	7.0	80.5	55.5	167.0
8	Б$_{80}$(3+2+1+1+1)	7.0	202.8	134.6	304.5
9	$N_{350}P_{250}K_{170}$+Б$_{40}$(40)	6.8	128.0	109.4	187.7
10	$N_{350}P_{250}K_{170}$+H_{90}(3+2+2+2)	6.6	173.7	56.4	180.1
2020 years					
1	$N_{350}P_{250}K_{170}$	6.7	37.6	37.5	67.3
2	$N_{350}P_{250}K_{170}$+Б$_{10}$(1)	6.8	79.7	68.4	108.7
3	$N_{350}P_{250}K_{170}$+Б$_{20}$(1+1)	6.9	74.1	71.2	125.8
4	$N_{350}P_{250}K_{170}$+Б$_{30}$(1+1+1)	6.8	126.3	103.3	166.7
5	$N_{350}P_{250}K_{170}$+Б$_{40}$(1+1+1+1)	6.8	145.7	119.3	188.7
6	$N_{350}P_{250}K_{170}$+Б$_{50}$) (1+1+1+1+1)	6.9	141.0	129.5	203.2
7	Б$_{50}$(2+1+1+1)	6.9	85.51	70.6	179.0
8	Б$_{80}$(3+2+1+1+1)	7.1	210.8	142.2	323.0
9	$N_{350}P_{250}K_{170}$+Б$_{40}$(40)	7.0	143.9	121.8	197.1
10	$N_{350}P_{250}K_{170}$+H_{90}(3+2+2+2)	6.7	160.5	73.7	150.7
2021 years					
1	$N_{350}P_{250}K_{170}$	6.7	41.8	55.5	93.4
2	$N_{350}P_{250}K_{170}$+Б$_{10}$(1)	6.7	91.1	74.8	122.0
3	$N_{350}P_{250}K_{170}$+Б$_{20}$(1+1)	6.7	105.5	81.8	131.5
4	$N_{350}P_{250}K_{170}$+Б$_{30}$(1+1+1)	6.8	135.8	109.9	182.5
5	$N_{350}P_{250}K_{170}$+Б$_{40}$(1+1+1+1)	6.9	165.6	126.6	205.8
6	$N_{350}P_{250}K_{170}$+Б$_{50}$) (1+1+1+1+1)	6.9	157.9	140.9	231.7
7	Б$_{50}$(2+1+1+1)	6.9	95.1	80.0	207.5
8	Б$_{80}$(3+2+1+1+1)	7.1	217.8	155.6	346.1
9	$N_{350}P_{250}K_{170}$+Б$_{40}$(40)	7.0	165.0	136.7	211.0
10	$N_{350}P_{250}K_{170}$+H_{90}(3+2+2+2)	6.6	162.0	74.1	200.2

Appendix 8.8 The Quantity of NPK Present in the Greenhouse Soil Before the Fourth Plant Feeding in the Period of Robust Plant Productivity

№	*Options*	*pH*	*The amount of NPK contained in the soil before fourth feeding, mg/kg*		
			Nм	P_2O_5	K_2O
2019 years					
1	$N_{350}P_{250}K_{170}$	6.8	14.3	23.2	19.3
2	$N_{350}P_{250}K_{170}$+$Б_{10}$(1)	6.7	29.2	44.3	47.5
3	$N_{350}P_{250}K_{170}$+$Б_{20}$(1+1)	6.7	24.7	36.1	30.6
4	$N_{350}P_{250}K_{170}$+$Б_{30}$(1+1+1)	6.7	51.7	55.6	42.8
5	$N_{350}P_{250}K_{170}$+$Б_{40}$(1+1+1+1)	6.8	37.8	56.3	37.9
6	$N_{350}P_{250}K_{170}$+$Б_{50}$) (1+1+1+1+1)	6.8	51.7	68.6	44.1
7	$Б_{50}$(2+1+1+1)	7.0	37.1	36.4	79.5
8	$Б_{80}$(3+2+1+1+1)	7.1	149.0	102.0	208.0
9	$N_{350}P_{250}K_{170}$+$Б_{40}$(40)	7.0	90.0	78.6	96.0
10	$N_{350}P_{250}K_{170}$+H_{90}(3+2+2+2)	6.7	107.0	36.3	73.1
2020 years					
1	$N_{350}P_{250}K_{170}$	6.6	10.9	22.7	15.8
2	$N_{350}P_{250}K_{170}$+$Б_{10}$(1)	6.7	33.1	49.9	55.1
3	$N_{350}P_{250}K_{170}$+$Б_{20}$(1+1)	6.7	34.9	42.4	47.6
4	$N_{350}P_{250}K_{170}$+$Б_{30}$(1+1+1)	6.9	59.1	66.4	54.0
5	$N_{350}P_{250}K_{170}$+$Б_{40}$(1+1+1+1)	6.9	64.6	73.5	47.6
6	$N_{350}P_{250}K_{170}$+$Б_{50}$) (1+1+1+1+1)	6.9	60.8	83.1	63.7
7	$Б_{50}$(2+1+1+1)	6.9	38.7	40.1	84.6
8	$Б_{80}$(3+2+1+1+1)	6.9	152.0	116.0	228.0
9	$N_{350}P_{250}K_{170}$+$Б_{40}$(40)	7.0	98.1	87.5	108.7
10	$N_{350}P_{250}K_{170}$+H_{90}(3+2+2+2)	6.6	94.5	32.1	56.0
2021 years					
1	$N_{350}P_{250}K_{170}$	6.8	23.5	30.2	26.7
2	$N_{350}P_{250}K_{170}$+$Б_{10}$(1)	6.8	38.0	55.6	72.2
3	$N_{350}P_{250}K_{170}$+$Б_{20}$(1+1)	6.9	44.7	57.5	53.1
4	$N_{350}P_{250}K_{170}$+$Б_{30}$(1+1+1)	6.8	70.0	73.7	70.7
5	$N_{350}P_{250}K_{170}$+$Б_{40}$(1+1+1+1)	6.9	73.6	77.4	73.1
6	$N_{350}P_{250}K_{170}$+$Б_{50}$) (1+1+1+1+1)	6.9	74.1	107.0	86.0
7	$Б_{50}$(2+1+1+1)	6.8	46.1	52.6	102.2
8	$Б_{80}$(3+2+1+1+1)	6.9	160.0	128.0	235.3
9	$N_{350}P_{250}K_{170}$+$Б_{40}$(40)	6.9	108.0	119.0	121.8
10	$N_{350}P_{250}K_{170}$+H_{90}(3+2+2+2)	6.6	91.6	40.4	77.7

Appendix 8.9 The Total NPK Content That Has Accrued in the Greenhouse Soil After the Fourth Plant Feeding, Within the Period of Vigorous Plant Productivity

№	*Options*	*pH*	*Amount of NPK accumulated in greenhouse soil after 4-feeding, mg/kg*		
			Nм	P_2O_5	K_2O
2019 years					
1	$N_{350}P_{250}K_{170}$	6.7	43.3	27.6	34.6
2	$N_{350}P_{250}K_{170}$+$Б_{10}$(1)	6.7	58.2	53.7	60.8
3	$N_{350}P_{250}K_{170}$+$Б_{20}$(1+1)	6.8	46.2	50.1	57.4
4	$N_{350}P_{250}K_{170}$+$Б_{30}$(1+1+1)	6.8	80.7	81.3	70.0
5	$N_{350}P_{250}K_{170}$+$Б_{40}$(1+1+1+1)	6.9	91.8	91.4	86.0
6	$N_{350}P_{250}K_{170}$+$Б_{50}$) (1+1+1+1+1)	6.9	111.0	89.4	137.2
7	$Б_{50}$(2+1+1+1)	7.0	52.1	47.1	124.3
8	$Б_{80}$(3+2+1+1+1)	7.1	185.0	124.0	276.7
9	$N_{350}P_{250}K_{170}$+$Б_{40}$(40)	7.0	116.0	98.0	179.3
10	$N_{350}P_{250}K_{170}$+H_{90}(3+2+2+2)	6.6	134.0	42.7	112.4
2020 years					
1	$N_{350}P_{250}K_{170}$	6.7	39.9	44.1	45.1
2	$N_{350}P_{250}K_{170}$+$Б_{10}$(1)	6.7	62.1	65.3	88.4
3	$N_{350}P_{250}K_{170}$+$Б_{20}$(1+1)	6.7	53.9	63.4	74.4
4	$N_{350}P_{250}K_{170}$+$Б_{30}$(1+1+1)	6.8	89.1	92.1	101.2
5	$N_{350}P_{250}K_{170}$+$Б_{40}$(1+1+1+1)	6.9	103.6	105.6	91.7
6	$N_{350}P_{250}K_{170}$+$Б_{50}$) (1+1+1+1+1)	6.9	119.8	130.0	156.8
7	$Б_{50}$(2+1+1+1)	7.0	68.7	60.8	129.4
8	$Б_{80}$(3+2+1+1+1)	7.1	186.0	138.0	250.7
9	$N_{350}P_{250}K_{170}$+$Б_{40}$(40)	7.0	123.1	116.9	122.0
10	$N_{350}P_{250}K_{170}$+H_{90}(3+2+2+2)	6.7	119.0	58.5	85.3
2021 years					
1	$N_{350}P_{250}K_{170}$	6.7	52.5	66.6	49.9
2	$N_{350}P_{250}K_{170}$+$Б_{10}$(1)	6.7	70.0	80.0	95.5
3	$N_{350}P_{250}K_{170}$+$Б_{20}$(1+1)	6.9	76.2	85.5	79.9
4	$N_{350}P_{250}K_{170}$+$Б_{30}$(1+1+1)	6.8	100.0	100.0	107.7
5	$N_{350}P_{250}K_{170}$+$Б_{40}$(1+1+1+1)	6.9	127.6	113.0	121.4
6	$N_{350}P_{250}K_{170}$+$Б_{50}$) (1+1+1+1+1)	6.9	140.2	148.0	178.9
7	$Б_{50}$(2+1+1+1)	7.0	76.1	78.3	147.0
8	$Б_{80}$(3+2+1+1+1)	7.1	195.6	150.0	273.9
9	$N_{350}P_{250}K_{170}$+$Б_{40}$(40)	7.0	140.0	128.0	145.2
10	$N_{350}P_{250}K_{170}$+H_{90}(3+2+2+2)	6.7	128.0	76.8	127.2

Appendix 8.10 The Results of the Analysis of Soil Composition After Harvesting the Entire Crop

№	*Options*	*pH*	*Amount of NPK in harvested soil, mg/kg*		
			Nм	P_2O_5	K_2O
2019 years					
1	$N_{350}P_{250}K_{170}$	6.7	18.9	23.0	25.6
2	$N_{350}P_{250}K_{170}$+$Б_{10}$(1)	6.7	27.9	41.5	30.2
3	$N_{350}P_{250}K_{170}$+$Б_{20}$(1+1)	6.8	34.9	34.3	20.3
4	$N_{350}P_{250}K_{170}$+$Б_{30}$(1+1+1)	6.8	45.5	56.2	29.7
5	$N_{350}P_{250}K_{170}$+$Б_{40}$(1+1+1+1)	6.9	50.8	52.8	36.4
6	$N_{350}P_{250}K_{170}$+$Б_{50}$) (1+1+1+1+1)	6.9	64.0	67.3	78.3
7	$Б_{50}$(2+1+1+1)	7.0	27.9	27.6	75.5
8	$Б_{80}$(3+2+1+1+1)	7.1	110.0	108.0	203.0
9	$N_{350}P_{250}K_{170}$+$Б_{40}$(40)	7.0	93.0	78.2	75.0
10	$N_{350}P_{250}K_{170}$+H_{90}(3+2+2+2)	6.8	91.0	39.9	78.3
2020 years					
1	$N_{350}P_{250}K_{170}$	6.7	16.3	29.5	22.9
2	$N_{350}P_{250}K_{170}$+$Б_{10}$(1)	6.7	40.4	50.5	45.0
3	$N_{350}P_{250}K_{170}$+$Б_{20}$(1+1)	6.9	35.4	49.8	24.8
4	$N_{350}P_{250}K_{170}$+$Б_{30}$(1+1+1)	6.8	63.4	75.0	35.9
5	$N_{350}P_{250}K_{170}$+$Б_{40}$(1+1+1+1)	6.9	76.2	87.0	49.0
6	$N_{350}P_{250}K_{170}$+$Б_{50}$) (1+1+1+1+1)	7.0	79.7	107.0	97.0
7	$Б_{50}$(2+1+1+1)	6.9	35.1	42.6	91.3
8	$Б_{80}$(3+2+1+1+1)	6.9	137.5	114.0	221.0
9	$N_{350}P_{250}K_{170}$+$Б_{40}$(40)	7.0	107.1	95.0	88.6
10	$N_{350}P_{250}K_{170}$+H_{90}(3+2+2+2)	6.8	90.1	44.4	34.1
2021 years					
1	$N_{350}P_{250}K_{170}$	6.8	20.2	45.9	31.2
2	$N_{350}P_{250}K_{170}$+$Б_{10}$(1)	6.9	50.0	63.1	72.1
3	$N_{350}P_{250}K_{170}$+$Б_{20}$(1+1)	6.9	48.6	65.3	41.9
4	$N_{350}P_{250}K_{170}$+$Б_{30}$(1+1+1)	6.8	75.1	90.2	48.8
5	$N_{350}P_{250}K_{170}$+$Б_{40}$(1+1+1+1)	6.9	90.0	101.0	66.3
6	$N_{350}P_{250}K_{170}$+$Б_{50}$) (1+1+1+1+1)	6.9	101.0	130.0	126.0
7	$Б_{50}$(2+1+1+1)	7.0	49.4	65.5	112.0
8	$Б_{80}$(3+2+1+1+1)	7.1	146.0	136.0	239.0
9	$N_{350}P_{250}K_{170}$+$Б_{40}$(40)	7.0	116.0	130.0	114.0
10	$N_{350}P_{250}K_{170}$+H_{90}(3+2+2+2)	6.7	100.0	61.8	94.6

Appendix 8.11 The Impact of Biohumus on the Concentration of Heavy Metals in the Soil While Cultivating Cucumbers Within a Greenhouse Environment

№	*The amount of biohumus applied to the soil is kg/m²*	*Amount of heavy metals in mobile form in harvested soil, mg/kg*								
		Cd	*Co*	*Cr*	*Cu*	*Mn*	*Ni*	*Pb*	*Zn*	*Mo*
2019 years										
1	$N_{350}P_{250}K_{170}$	0.34	0.39	1.51	2.82	45.80	0.87	5.07	46.00	0.16
2	$N_{350}P_{250}K_{170}$+Б$_{10}$(1)	0.31	0.27	1.44	2.51	42.61	0.78	3.41	31.11	0.16
3	$N_{350}P_{250}K_{170}$+Б$_{20}$(1+1)	0.20	0.21	1.32	2.26	36.87	0.78	2.03	22.11	0.15
4	$N_{350}P_{250}K_{170}$+Б$_{30}$(1+1+1)	0.10	0.14	1.12	1.21	29.61	0.61	1.21	14.23	0.13
5	$N_{350}P_{250}K_{170}$+Б$_{40}$(1+1+1+1)	0.09	0.13	1.10	1.36	29.57	0.66	1.41	15.41	0.12
6	$N_{350}P_{250}K_{170}$+Б$_{50}$(1+1+1+1+1)	0.09	0.15	1.10	1.00	29.70	0.60	1.61	13.73	0.12
7	Б$_{50}$(2+1+1+1)	0.08	0.10	0.83	0.79	23.40	0.58	1.23	12.83	0.11
8	Б$_{80}$(3+2+1+1+1)	0.08	0.12	0.87	0.72	23.52	0.55	1.11	11.20	0.10
9	$N_{350}P_{250}K_{170}$+Б$_{40}$(40)	0.06	0.08	0.63	0.70	18.90	0.38	0.73	9.30	0.08
10	$N_{350}P_{250}K_{170}$+H$_{90}$(3+2+2+2)	0.30	0.32	1.30	2.36	37.5	0.77	3.40	33.17	0.15
2020 years										
1	$N_{350}P_{250}K_{170}$	0.42	0.50	1.59	2.91	46.8	0.81	4.86	45.13	0.17
2	$N_{350}P_{250}K_{170}$+Б$_{10}$(1)	0.33	0.23	1.39	2.46	42.6	0.75	3.16	35.21	0.15
3	$N_{350}P_{250}K_{170}$+Б$_{20}$(1+1)	0.18	0.18	1.29	2.44	37.8	0.77	2.28	24.21	0.13
4	$N_{350}P_{250}K_{170}$+Б$_{30}$(1+1+1)	0.09	0.11	1.10	0.98	29.0	0.58	1.39	13.33	0.12
5	$N_{350}P_{250}K_{170}$+Б$_{40}$(1+1+1+1)	0.08	0.18	1.05	1.14	29.5	0.65	1.01	12.51	0.11
6	$N_{350}P_{250}K_{170}$+Б50(1+1+1+1+1)	0.06	0.11	1.05	0.91	29.1	0.57	1.16	11.83	0.10
7	Б$_{50}$(2+1+1+1)	0.08	0.07	0.80	0.96	23.4	0.56	1.08	8.93	0.09
8	Б$_{80}$(3+2+1+1+1)	0.06	0.08	0.84	0.61	22.5	0.54	1.06	9.34	0.08
9	$N_{350}P_{250}K_{170}$+Б$_{40}$(40)	0.05	0.06	0.59	0.58	18.9	0.35	0.78	7.45	0.06
10	$N_{350}P_{250}K_{170}$+H$_{90}$(3+2+2+2)	0.38	0.37	1.51	2.63	43.0	0.70	3.58	37.24	0.20
2021 years										
1	$N_{350}P_{250}K_{170}$	0.38	0.46	1.48	2.98	47.2	0.85	4.75	41.97	0.14
2	$N_{350}P_{250}K_{170}$+Б$_{10}$(1)	0.31	0.24	1.37	2.49	40.5	0.76	3.19	32.01	0.14
3	$N_{350}P_{250}K_{170}$+Б$_{20}$(1+1)	0.17	0.17	1.26	2.26	34.3	0.75	2.39	19.98	0.13
4	$N_{350}P_{250}K_{170}$+Б$_{30}$(1+1+1)	0.10	0.13	1.08	1.09	27.0	0.59	1.19	11.93	0.10
5	$N_{350}P_{250}K_{170}$+Б$_{40}$(1+1+1+1)	0.06	0.15	1.05	1.25	27.0	0.64	1.10	10.81	0.10
6	$N_{350}P_{250}K_{170}$+Б$_{50}$(1+1+1+1+1)	0.08	0.10	1.03	1.18	27.1	0.58	1.29	11.63	0.11
7	Б$_{50}$(2+1+1+1)	0.06	0.07	0.78	0.87	20.8	0.56	1.07	8.03	0.10
8	Б$_{80}$(3+2+1+1+1)	0.06	0.09	0.82	0.71	20.9	0.53	0.90	7.81	0.09
9	$N_{350}P_{250}K_{170}$+Б$_{40}$(40)	0.06	0.08	0.57	0.69	16.3	0.36	0.61	7.28	0.07
10	$N_{350}P_{250}K_{170}$+H$_{90}$(3+2+2+2)	0.31	0.33	1.51	2.39	43.9	0.77	3.49	40.10	0.19

REFERENCES

Alekseev Y.V. *Heavy Metals in Soils and Plants*. Leningrad: Agropromizdat, Leningradskoe otd-e, 1987, p. 142 (in Russian).

Bairagi S.K., Singh D.K., Ram H.H. Analysis of combining ability in cucumber (Cucumis sativus L.) through half dialled mating system. *Annals of Horticulture*, 2013, 6(2): 308–314.

Bryzgalov V.A. *Vegetable Growing of Protected Ground: [According to Special 1503 "Fruit and Vegetable Growing and Viticulture"]*, edited by V.A. Bryzgalov, V.E. Sovetkina, N.I. Savinova. Leningrad: Leningr. Compartment, 1983, 352 p.

Dospekhov B.A. *Methodology of Field Experience*. Moscow: Agropromizdat, 1985, p. 351.

Egel D.S. *Midwest Vegetable Production Guide for Commercial Growers*. University of Illinois Extension, Purdue Extension, Iowa State Extension, Kansas State Extension, University of Minnesota Extension, University of Missouri Extension, 2015, p. 210.

Farhan M., Mastura M.T., Ansari S.P., Muaz M., Azeem M., Sapuan S.M. Advanced potential hybrid biocomposites in aerospace applications: a comprehensive review. In *Advanced Composites in Aerospace Engineering Applications*. Cham: Springer Nature, Switzerland AG, 2022. https://doi.org/10.1007/978-3-030-88192-4_6.

Farhan M., Sadique M., Shadab Khan M. *Design &Development of Small Portable Table by Sugar Palm Fibre Reinforced Unsaturated Polyester Composites*. National Conference Synthesis, Properties and Application (NCFM-2022) organized by Department of Physics, Aligarh Muslim University, Aligarh, India.

Farhan M., Shafiq Ansari M. *Engineering Efficacy of Sugar Palm Tree Fibre (Arenga pinnata) Grown in Various Regions of India: A Review*. ICCIE-2019: 2nd International Conference on Chemistry, Industry and Environment organized by Department of Applied Chemistry, Z.H. College of Engineering & Technology, Aligarh Muslim University, Aligarh, (India), 2019.

Fernandes A., Martinez H.E.P., Oliveira L.R. Effect of nutrient sources on yield, fruit quality and nutritional status of cucumber plants cultivated in hydroponics. *Horticultura Brasileira*, 2002, 20(4): 571–575.

Gluntsov N.M., Vendilo G.G., Ovcharenko M.M. et al. *Guidelines for the Determination of Heavy Metals in Greenhouse Soil and Vegetable Products*. Moscow, 1996, p. 19.

Ilyin V.B. *Heavy Metals in the Soil-Plant System*. Novosibirsk: Nauka, Sib. otd-e, 1991, p. 151 (in Russian).

Imran M.F., Akhtar J., Sumita M.Z. *Recent Developments in Single-Use Plastic Packaging Sustainable Alternatives in Food Industry: An Affordable Way to Go Economic Green*. INDO-UZBEK MEET & International Conference on Trends & Innovations in Food Technology from Farm to Fork (TIFT-2022), organized by the dept. of Bioengineering Integral University, Lucknow & co-organized by (TCTI), Uzbekistan, 2022.

Janapriya S., Palanisamy D., Ranghaswami M.V. Soilless media and fertigation for naturally ventilated polyhouse production of cucumber (Cucumis sativus) cv. green long. *International Journal of Agriculture Environment and Biotechnology*, 2010, 3(2): 199–205.

Karpova E.A. *State of Trace Elements in Agroecosystems: Tr. Biogeochemical Laboratory*. Moscow: Nauka, 2003, T.24. S. 6–87.

Kasatikova S.M. Tests of vermicompost/S.M. Kasatikova, V.A. Kasatikov. *Agrochemical Bulletin*, 2002, 6: 29–30.

Khujamshukurov N., Eshkobilov Sh.A., Kuchkarova D.X., Bashirova Yu.J. (a) *Influence of Biohumus on the Content of Heavy Metals in Soil.* International Conference on Trends & Innovations in Food Technology. 24–25 November 2022 (TIFT-2022). Integral University, Lucknow, India, p. 85.

Khujamshukurov N., Eshkobilov Sh.A., Normatov A., Kuchkarova D.X., Bashirova Yu.J. (b) *Influence of Biohumus on the Productivity of Cucumber Plants Under Greenhouse Conditions.* International Conference on Trends & Innovations in Food Technology. 24–25 November, 2022 (TIFT-2022). Integral University, Lucknow, India. p. 86.

Krug H., Liebig H.P., Stützel H. *Gemüseproduktion.* Stuttgart: Eugen Ulmer Gmb H & Co, 2002, p. 463.

Kubota Ch, Balliu A, Nicola, S. Quality of planting material. In *Good Agricultural Practices for Greenhouse Vegetable Crops. Principles for Mediterranean Climate Areas.* Rome: FAO, Plant Production and Protection Paper 217, 2013, pp. 355–378.

Kutukova Y.D. *The State of Heavy Metals in Soils and Their Accumulation in Plants During the Introduction of Sewage Sludge and Ameliorants.dis . . . . k. b. n. M*, 2001, p. 25 (in Russian).

Kutukova Yu.D., Plekhanova I.O. *Influence of Ameliorants on the State of Heavy Metals in Soils and Their Content in Plants When Using Sewage Sludge as a Fertilizer*, 2002, 12, pp. 68–74.

Liu F., Fu X., Wu G., Feng Y., Li F., Bi H., Ai X. Hydrogen peroxide is involved in hydrogen sulfide-induced carbon assimilation and photoprotection in cucumber seedlings. *Environmental and Experimental Botany*, 2020, 175, p. 104052. https://doi.org/10.1016/j.envexpbot.2020.104052.

El-Gamal A., Saleh I. (2016). Geochemical assessment of heavy metals pollution and ecological risk in the Nile delta coastal sediments. *Egypt*, 2016, 26: 41–59. https://doi.org/10.4197/Mar.26-1.5.

Obukhov A.I., Efremova L.L. *Protection and Reclamation of Soils Contaminated with Heavy Metals.* Proceedings of the 2nd All-Union Conference "Heavy Metals in the Environment and Nature Conservation." Moscow, 1988. 4.1. S.23–35.

Obukhov A.I., Plekhanova I.O. Detoxification of sod-podzolic soils contaminated with heavy metals: Theoretical and practical aspects, 1995, 2: 108–116.

Obukhov A.I., Popova A.A. Balance of heavy metals in agrocenoses of sod-podzolic soils and monitoring problems. Bulletin of Moscow University, ser. 17. *Soil Science*, 1992, 3. S. 31–39.

Pal A., Adhikary R., Shankar T., Sahu A.K., Maitra S. Cultivation of cucumber in greenhouse. In *Protected Cultivation and Smart Agriculture.* New Delhi: New Delhi Publishers, 2020, pp. 139–145.

Plekhanova I.O., Kutukova Yu.D., Obukhov A.I. Accumulation of heavy metals by agricultural plants during the introduction of sewage sludge. *Soil Science*, 1995a, 12: S.1530–1536.

Plekhanova I.O., Kutukova Yu.D., Obukhov A.I. Accumulation of heavy metals by agricultural plants when introducing sewage sediments. *Eurasian Soil Science*, 1995b, 12: 1530–1535.

Plekhanova I.O., Savelyeva V.A. Influence of ameliorants on the state of cobalt in the soil and its entry into plants. *Agrochemistry*, 1997, 8: 68–73.

Savvas D., Gianquinto G., Tüzel Y., Gruda N. Soilless culture. In *Good Agricultural Practices for Greenhouse Vegetable Crops. Principles for Mediterranean Climate Areas.* Rome: FAO, Plant Production and Protection Paper 217, 2013, pp. 303–354.

Sharma D., Sharma V.K., Kumari, A. Effect of spacing and training on growth and yield of polyhouse grown hybrid cucumber (Cucumis sativus L.). *International Journal of Current Microbiology and Applied Sciences*, 2018, 7(5): 1844–1852.

Singh B. *Protected Cultivation of Vegetable Crops.* Ludhiana, India: Kalyani Publishers, 2005, pp. 74–84.

Taha N., Abdalla N., Bayoumi Y., El-Ramady H. Management of greenhouse cucumber production under arid environments: A review. *Environment Biodiversity Soil Security*, 2020, 4: 123–136.

Voitovich N.V., Polev N.A., Ostanina A.V. *Assessment of Soil Pollution of Agricultural Use as a Result of Agrogenic Impact. Soils of the Moscow Region and Their Use.* Moscow: Soils. V.V. Dokuchaev Institute, 2002. T.1., pp. S.372–384.

Chapter 9

Smart ergonomic design of VDT workstation

An Affordable Way to Environment Health and Safety

Muhammad Farhan, Shara Khursheed, Mohammad Azad Alam, Mohammad Azeem, and Muhammed Muaz

9.1 INTRODUCTION

Anthropometric measurements have been an essential part of human history, dating back to ancient times. Anthropometry is a scientific discipline that involves the measurement of the human body to study human anatomy, physiology, and behavior. In this chapter, we will explore the importance of anthropometric measurements, anthropometric techniques and tools, body composition analysis, interpretation of anthropometric measurements, applications of anthropometric measurements, and the challenges and limitations of this field.

Wheelchair users typically have to use their wheelchairs solely for daily and professional tasks. As a result, people who use wheelchairs should be regarded as an intrinsic part of those chairs. Jobs involving visual display terminals (VDTs) are becoming more prevalent in contemporary workplaces, and wheelchair users frequently carry out these jobs owing to their nature of work (Jarosz, 1996). However, when VDT activities are conducted by wheelchair users, contact among those who operate (with the chairs) and workstation components is a significant concern. The employability of the potential employee, productivity, and danger of injury are all considerably harmed in the workplace for those who cannot "fit" into the workstation (Babski-Reeves et al., 2005; Kozey and Das 2004; Park et al., 2000). Ashworth et al. (1994) thought about how much older workers and individuals with disabilities are "planned out" of workplaces. Interesting research subjects include the appropriateness of current VDT workstation designs and the best or acceptable design for wheelchair users. Recently, there has been a great deal of research on VDT workstations, with the majority of these studies concentrating on occupational dangers such as felt weariness, pain with the eyes, and musculoskeletal strains for typical VDT operators. Anthropometric assessments and ergonomic concerns for people with physical disabilities, such as wheelchair users, have both been researched (Porter et al., 2004).

DOI: 10.1201/9781003495314-9

9.1.1 Anthropometry as a scientific discipline

Anthropometry is a scientific discipline that studies the human body's physical dimensions and proportions. This field involves the collection and analysis of body measurements, including height, weight, and various body dimensions. Anthropometry provides quantitative data that can be used to analyze the physical characteristics of humans and their evolution over time.

Anthropometry has been used for various purposes throughout history. The ancient Egyptians used anthropometry to construct the pyramids, while the Greeks used it to study the human form and the ideal proportions of the body. In the modern era, anthropometry has become a crucial tool for assessing human health, growth, and development. Anthropometry is used in various fields, including medicine, sports science, ergonomics, and anthropology (Psihogios et al., 2001; Reed and Roosmalan, 2005).

9.1.2 The importance of anthropometric measurements

Anthropometric measurements are essential for various reasons. They provide valuable information on an individual's physical characteristics, which can be used to monitor growth and development, assess nutritional status, and evaluate the effectiveness of interventions aimed at improving health. Anthropometric measurements can also identify individuals who are at risk for various diseases, such as obesity, diabetes, and cardiovascular disease. In addition, these measurements can help researchers and healthcare professionals develop and implement effective public health policies (Robertson et al., 2013).

9.1.3 Anthropometric techniques

Anthropometric measurements can be obtained using various techniques. Direct measurement techniques involve measuring the body directly using specialized tools such as tape measures and calipers. Indirect measurement techniques involve estimating body dimensions using mathematical equations or predictive models. Some commonly used techniques for anthropometric measurements include the following:

Height measurement: Height is typically measured using a stadiometer or a wall-mounted tape measure. The individual stands barefoot with their back straight, and their height is measured to the nearest millimeter.

Weight measurement: Weight is measured using a calibrated scale. The individual stands on the scale with minimal clothing and no shoes, and their weight is measured to the nearest kilogram.

Body mass index (BMI) calculation: BMI is calculated by dividing an individual's weight in kilograms by their height in meters squared. BMI is used to assess whether an individual is underweight, normal weight, overweight, or obese (Sengupta and Das, 1997).

Waist circumference measurement: Waist circumference is measured using a tape measure placed around the narrowest part of the waist. Waist circumference is used to assess the risk of obesity-related diseases such as diabetes and cardiovascular disease (Sundin et al., 2000a).

Hip circumference measurement: Hip circumference is measured using a tape measure placed around the widest part of the hips. Hip circumference is used to assess the risk of obesity-related diseases such as diabetes and cardiovascular disease (Sundin et al., 2000b).

Skin-fold thickness measurement: Skin-fold thickness is measured using calipers to measure the thickness of the subcutaneous fat layer at various sites on the body. Skin-fold thickness is used to estimate body fat percentage (Tayyari and Smith, 1997).

Bioelectrical impedance analysis (BIA): BIA is a non-invasive method for estimating body composition. BIA measures the resistance of body tissues to an electrical current and uses this measurement to estimate body fat percentage (McAtamney and Corlett, 1993).

9.1.4 Measurement tools

Anthropometric measurements require specialized tools that are calibrated and maintained regularly to ensure accuracy. These tools include the following:

1. Stadiometers: Used to measure height accurately.
2. Scales: Used to measure weight accurately.
3. Tape Measures: Used to measure waist circumference, hip circumference, and skinfold thickness.
4. Calipers: Used to measure skin fold thickness and estimate body fat percentage.
5. BIA devices: This is a non-invasive method for estimating the body composition. BIA measures the resistance of body tissues to an electrical current and uses this measurement to estimate body fat percentage (Tilley, 2002).

Standardization of measurement techniques is essential to ensure consistency across different individuals and populations. To achieve this, standardized protocols have been developed for each measurement technique, specifying the equipment to be used, the position of the individual, and the measurement procedure (Mattila, 1996).

9.1.5 Body composition analysis

Body composition analysis is the measurement of body fat and lean mass. Anthropometric measurements, such as skin fold thickness measurements and BIA, can be used to estimate body fat percentage. The World Health

Organization (WHO) has established guidelines for healthy body fat percentages, which vary by age and gender.

Body composition analysis is important for assessing the risk of obesity-related diseases such as diabetes and cardiovascular disease. It can also be used to evaluate the effectiveness of interventions aimed at reducing body fat, such as diet and exercise programs (Paquet and Feathers, 2004).

9.1.6 Interpretation of anthropometric measurements

Anthropometric measurements can provide valuable information on an individual's physical characteristics. The interpretation of these measurements depends on the specific measurement technique used and the population being studied. For example, BMI is used to assess whether an individual is underweight, normal weight, overweight, or obese, but its interpretation may differ for different populations.

Interpretation of anthropometric measurements must also take into account other factors, such as age, gender, and ethnicity. For example, some populations may have a higher prevalence of obesity-related diseases despite having a lower BMI. Ergonomic principles provide possibilities for optimizing tasks in the workplace. These principles are summarized in Table 9.1.

9.1.7 Applications of anthropometric measurements

Anthropometric measurements have many applications in various fields, including the following:

1. Medicine: Anthropometric measurements are used to monitor growth and development, assess nutritional status, and evaluate the risk of obesity-related diseases.
2. Sports science: Anthropometric measurements are used to evaluate athletic performance and design training programs.
3. Ergonomics: Anthropometric measurements are used to design ergonomic products and workstations that fit the physical characteristics of the user.
4. Anthropology: Anthropometric measurements are used to study the physical characteristics of different populations and their evolution over time.
5. Public Health: Anthropometric measurements are used to develop and implement effective public health policies aimed at reducing the prevalence of obesity-related diseases (Woodson et al., 1992; Wang and Verriest, 1998).

Table 9.1 Ergonomic Principles Provide Possibilities for Optimizing Tasks in the Workplace

S. No.	*Ergonomic principle*	*Different posture*	*Description*
1.	Avoid Bending Forward		The upper part of the body of an adult weighs about 40kg on average. The further the trunk is bent forwards, the harder it is for the muscles and ligaments of the back to maintain the upper body in balance
2.	Limit the weight of a load that is lifted		There are guidance weight limits for both males and females detailed in Figure 2 of this document.
3.	Use mechanical aids	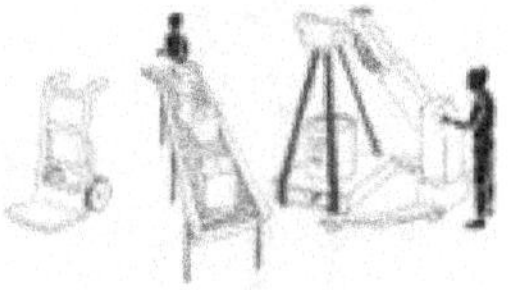	Many lifting accessories are available to help lift and move loads.
4.	Avoid carrying loads with one hand		When only one hand is used to carry a load, the body is subject to mechanical stress
5.	Use transport accessories	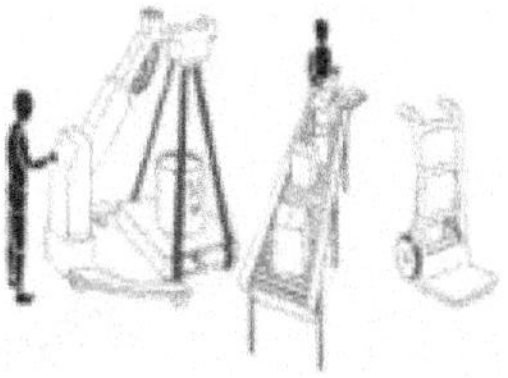	There are a large number of accessories such as roller conveyors, conveyor belts, trolleys and mobile raising platforms, which eliminate or reduce manual handling.

9.1.8 Challenges and limitations of anthropometric measurements

Anthropometric measurements have several challenges and limitations, including:

1. Inaccuracy: Anthropometric measurements may be inaccurate due to human error, instrument error, or measurement variability.
2. Variation: Anthropometric measurements may vary due to factors such as age, gender, and ethnicity.
3. Cost: Some anthropometric measurement tools, such as BIA devices, can be expensive.
4. Privacy: Anthropometric measurements may be considered invasive and may raise privacy concerns.
5. Cultural Sensitivity: Anthropometric measurements may be culturally sensitive and may not be accepted by all populations.

9.1.8.1 Common workplace postures

A number of ergonomic factors are taken into consideration while designing a workplace. They are work postures and movements, such as sitting, moving, pushing, standing, and lifting. Body postures impact work efficiency and human health. A bad work posture not only leads to early fatigue but also various musculoskeletal disorders shown in Figure 9.1. Therefore, an ergonomic workplace should be designed by the application of ergonomics principles of body postures, which play a key role in workplace design and safety. Body muscles, bones, and joints all contribute to workplace postures. Since muscles can support up to 40% of bodyweight, they play an essential role in tasks involving lifting and holding. Non-alignment in muscle movement without any support can become the cause of developing local mechanical stresses on the body parts like wrists, joints, shoulder, and neck. During the design of the workplace, the effect of gravity and anthropometric factors should be taken into consideration. Some common workplace postures are shown in Figure 9.3.

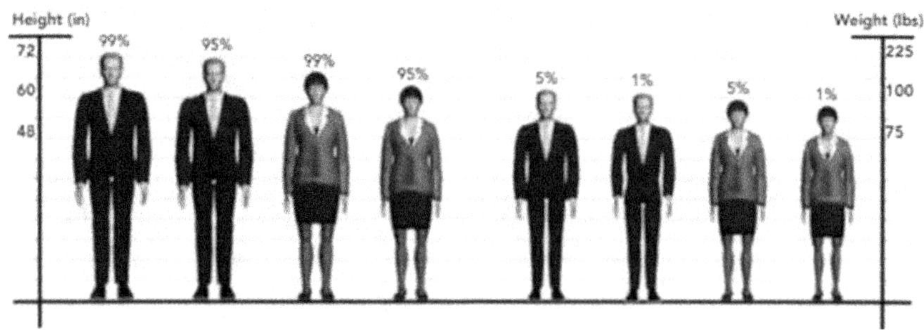

Figure 9.1 Numerous Relative Sizes of Human's Percentile (Wang et al., 1999).

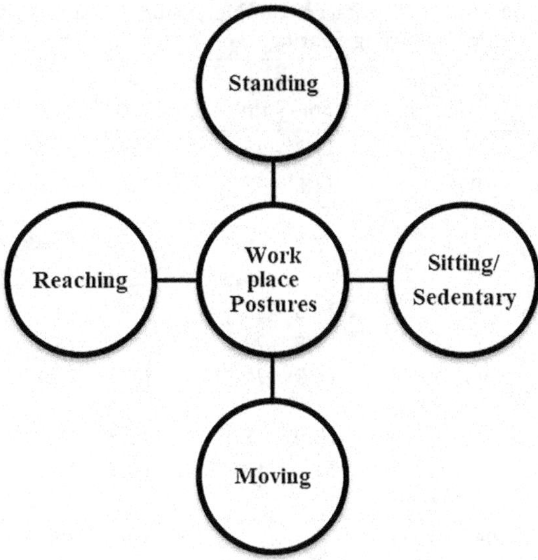

Figure 9.2 Various Types of Common Workplace Postures.

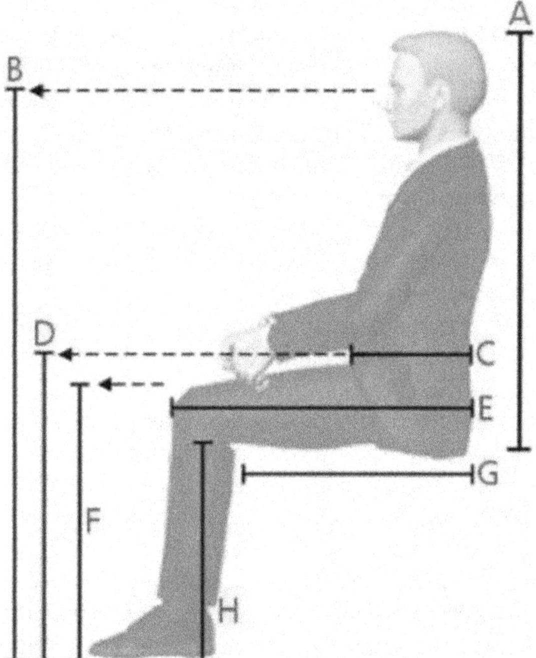

Figure 9.3 Common Anthropometric Measurements for the Seated Position. Use Table 9.2 for values.

Table 9.2 Value for 5th to 95th Percentile Males and Females in the Seated Position Used in Designing Seating. Use Figure 9.3 for visualization

Measurement	*Letter*	*Female 5th–95th%*	*Male 5th–95th%*	*Overall range 5th–95th%*
Sitting Height	A	31.3″–35.8″	33.6″–38.3″	31.3″–38.3″
Sitting Eye Height	B	42.6″–48.8″	46.3″–52.6″	42.6″–52.6″
Waist Depth	C	7.3″–10.7″	7.8″–11.4″	7.3″–11.4″
Thigh Clearance	D	21.0″–24.5″	23.0″–26.8″	21.0″–26.8″
Buttock-to-Knee	E	21.3″–25.2″	22.4″–26.3″	21.3–26.3″
Knee Height	F	19.8″–23.2″	21.4″–25.0″	19.8″–28.0″
Seat Length/ Depth	G	16.9″–20.4″	17.7″–21.1″	16.9″–21.1″
Popliteal Height	H	15.0″–18.1″	16.7″–19.9″	15.0″–19.9″
Seat Width	Not shown	14.5″–18.0″	13.9″–17.2″	13.9″–18.0″

Source: Data from BIFMA Ergonomics Guidelines, 2002. All measurements are in inches.

- *Standing*—The handling of loads during transportation, lifting of loads, and modern computer workstations are standing workplaces. A neutral position of joints and muscles is desirable while designing them. The outstretching of arms should be avoided by keeping the work piece close to the body (40–45 cms) (Chi, 1998). A desk height is taken as the parameter while designing workstations, and it is usually measured from the floor. Desk height can be different from the working height sometimes (Das and Kozey, 1999). The range of working height is 25″–48″ with an adjustable feature, which can be set according to the type of work, whether it is precision or heavy work. In the case of carrying the load while standing, both feet of the individual should be completely supported by the floor in an erect posture (Feyen et al., 2000; Dorian and David, 2023).
- *Sedentary Position*—A general problem developed during sedentary work is of back and neck pain. An adjustable table height is desirable depending on the anthropometric data (Dunstan et al., 2010; Farhan et al., 2022). The chair height should be design for 90–100 degree knee angle. Outstretching should be avoided, and the back should be properly supported. A footrest can be used to obtain a comfortable posture (Feney, 2000; Fogleman and Lewis, 2002)
- *Reaching*—The person, whether standing or sitting, might have to grasp tools or other things needed while working (Gerr et al., 2004; Guan et al., 2001). A particular arrangement of the table has been recommended for the same. Reach dimensions are calculated depending on the gender, weight, height, and so on. Separate places for regular and rare working zones are provided in the table. The working height

should be 50 cm below the elbow (Hedge, 1982). Workstations and accessories like pedestals and overhead storage should let the majority of the user's body joints to move within healthy ranges. The extent users must reach when developing items to reduce uncomfortable or harmful situations must be taken into consideration.

- *Moving*—Sometimes a person has to move in the workplace. The two types of movements preferred are parallel and symmetrical motions. One most important principle while working is use gravity doesn't oppose it (Gyi et al., 2004).

Parallel motion is preferred over the symmetrical motion but it not only gives easy shoulder move but also established better eye control (Murata et al., 2005).

9.2 COMMON WORKPLACE MOTIONS

The workplace should, in the end, be welcoming to users and as flexible as feasible. It is possible to increase worker productivity and reduce the risk of illness and injury by using workplace goods that are built with this in mind. The range of motion in the human body is inherent (ROM). Movement within the appropriate range of motion (ROM) encourages blood flow and flexibility, which may result in greater comfort and increased productivity (Lin et al., 2016). Despite the necessity to encourage mobility, users should aim to avoid monotonous motions and specific ROM extremes over extended periods of time. Products may be developed to function within the best ranges to assist decrease the incidence of tiredness and muscular diseases by taking into account both range of motion and repetitive action (Löffler et al., 2015).

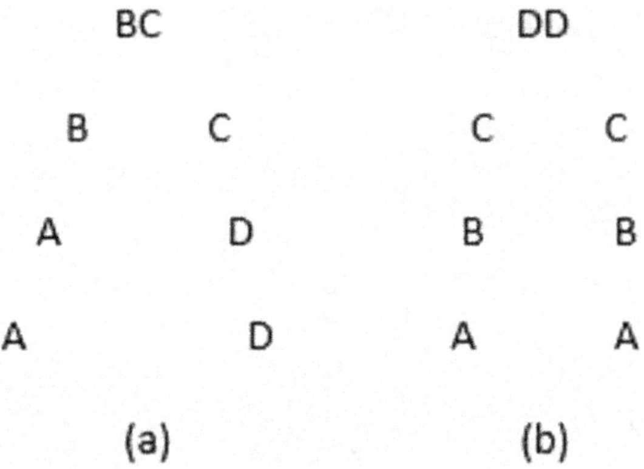

Figure 9.4 (a) Parallel motion; (b) symmetrical motion.

9.2.1 Good and bad zones

A user may come into contact with one of four separate zones when seated or standing: For the majority of movements, Zone 0 as well (Green Zone) is preferred. minimizes strain on joints and muscles. For the majority of moves, Zone 1 (Yellow Zones) is preferred. Minimizes strain on joints and muscles. Zone 2 (the Red Zone) places the limbs in a more severe posture and puts more strain on the muscles and joints. Zone 3 (beyond the red zone): Avoid using your limbs in the most extreme postures, especially while lifting large objects or performing repetitive motions. These are the areas where bodily parts are free to move. While Zone 2 and 3 depict more severe situations, Zone 0 and 1 reflect lesser joint motions (Haijanen et al., 1996; Leskinen and Haijanen, 1996; Leung et al., 2004.).

9.3 CONCLUSION AND GENERAL RECOMMENDATIONS

Anthropometric measurements are a critical tool in understanding human physical characteristics, assessing health status, and designing interventions for improving health outcomes. The measurement tools used in anthropometry must be calibrated and standardized to ensure accuracy and consistency across different populations. Body composition analysis, interpretation of anthropometric measurements, and applications of anthropometric measurements are important aspects of this discipline that have a range of applications in various fields. However, anthropometric measurements also face challenges and limitations such as inaccuracy, variation, cost, privacy, and cultural sensitivity. Despite these challenges, the benefits of anthropometric measurements far outweigh the limitations, making it a vital tool for researchers, healthcare providers, and policymakers to promote healthy living and reduce the prevalence of obesity-related diseases.

9.3.1 Screen setting

a letter size of 3.4 mm in height, a moderate screen distance of 710–930 mm, and a viewing angle of 0–15.

9.3.2 Postures

The majority of drivers favored positions they take when driving. Backbend (104–113 degrees); elbows slightly open (99 degrees).

9.3.3 Furniture

Keyboard height (from floor to the home row) is between 700 and 850, screen center is between 900 and 1150, screen is inclined to the horizontal by 88 to 105 degrees, key board home row is between 100 and 260 degrees, and screen is between 500 and 750 degrees.

9.3.4 Requirements of a VDT workstation

Height and distance from the screen are easily adjusted. At knee level, the distance from the table's front edge to the rear wall shouldn't be below 600 mm, and it should be a minimum of 800 mm when doing the feat.

9.3.5 Design of keyboards

The layout of keyboards' front height shouldn't be less than 20 mm, and the home row should be 30 mm above the desktop with a 5–15 degree incline: 17–19 mm between key centers, 0.4–0.8 N of force to activate the keys, and 3–5 mm of key displacement. Keyboards should be such that they can be moved as needed, with support for the arms and tactile input (feeling of acceptance). A split keyboard is currently being developed using ergonomic concepts. The two sections are designed with horizontal slopes of 10 to lessen the degree of inward bending of the arms and wrists and an opening inclination of 25 to prevent lateral twisting of the hands.

ACKNOWLEDGMENT

The authors thank Integral University, India, and Universiti Putra Malaysia for providing the facilities to carry out this research with technical support in writing this chapter. The authors are grateful to Prof. Sapuan Salit of (UPM), Malaysia, for guidance throughout the chapter. The authors also thank Dr. Mohammad Azeem of Universiti Technologi Petronas, Malaysia, for their advice and fruitful discussions.

REFERENCES

Ashworth, J. B., Reuben, D. B., and Benton, L. A. 1994. Functional profile of healthy older persons. *Age and Aging*, 23: 34–39.

Babski-Reeves, K., Stanfield, J., and Hughes, L. 2005. Assessment of video display workstation set up on risk factors associated with the development of low back and neck discomfort. *International Journal of Industrial Ergonomics*, 35: 593–604.

Chi, C. F. 1998. *The Task Redesign Manual for Handicapped People*. Bureau of Employment and Vocational Training, Council of Labor Affairs, Executive Yuan. (In Chinese)

Das, B., and Kozey, J. W. 1999. Structural anthropometric measurements for wheelchair mobile adults. *Applied Ergonomics*, 30: 385–390.

Dorian, S. T. E. F., and David, M. 2023. Ergonomics and postural assessment of dentists to prevent musculoskeletal disorders, designing an ergonomic chair. *Acta Technica Napocensis-Series: Applied Mathematics, Mechanics, and Engineering*, 65(3s).

Dunstan, D. W., et al. 2010. Television viewing time and mortality: The Australian diabetes, obesity and lifestyle study (AusDiab). *Circulation*, 121(3): 384–391.

Farhan, M., Mastura, M. T., Ansari, S. P., Muaz, M., Azeem, M., and Sapuan, S.M. 2022. Advanced potential hybrid biocomposites in aerospace applications: A comprehensive review. In *Advanced Composites in Aerospace Engineering Applications*. Cham: Springer Nature. https://doi.org/10.1007/978-3-030-88192-4_6.

Feeney, R. 2000. Approach with special ergonomic study of VDT workstations for wheelchair users. *International Journal of Applied Science Engineering*, 5(2): 113. Reference to the needs of the disabled person. Proceedings of the IEA 2000/HFES 2000 Congress.

Feyen, R., Liu, Y., Chaffin, D., Jimmerson, G., and Joseph, B. 2000. Computer-aided ergonomics: A case study of incorporating ergonomics analysis into workplace design. *Applied Ergonomics*, 31: 291–300.

Fogleman, M., and Lewis, R. J. 2002. Factors associated with self-reported musculoskeletal discomfort in video display terminal (VDT) users. *International Journal of Industrial Ergonomics*, 29: 311–318.

Gerr, F., Marcus, M., and Monteilh, C. 2004. Epidemiology of musculoskeletal disorders among computer users: Lesson learned from the role of posture and keyboard use. *Journal of Electromyography and Kinesiology*, 14: 25–31.

Guan, S. S., Lin, Y. C., and Ke, L. T. 2001. *A Study of Computer Workstation of Ergonomics*. Proceedings of 8th Annual Meeting & Conference of Ergonomics Society of Taiwan, 229–234. (In Chinese).

Gyi, D. E., Sims, R. E., Porter, J. M., Marshall, R., and Case, K. 2004, Representing older and disabled people in virtual user trials: Data collection methods. *Applied Ergonomics*, 35: 443–451.

Haijanen, J., Leskinen, T., Kuusisto, A., and Laitinen, H. 1996. Redesign of lifting work using a 3-D human modelling software and the revised NIOSH lifting equation. *Advances in Occupational Ergonomics and Safety, Finland*, 339–344.

Hedge, A. 1982. The open-plan office: A systematic investigation of employee reactions to their work environment. *Environment and Behavior*, 14(5): 519–542.

Jarosz, E. 1996.Determination of the workspace of wheelchair users. *International Journal of Industrial Ergonomics*, 17: 123–133.

Kozey, J. W., and Das, B. 2004. Determination of the normal and maximum reach measures of adult wheelchair users. *International Journal of Industrial Ergonomics*, 33: 205–213.

Leskinen, T., and Haijanen, J. 1996.*Torque on the Low Back and the Weight Limits Recommended by NIOSH in Simulated Lifts*. Proceedings of the Fourth International Symposium on 3-D Analysis of Human Movement, France.

Leung, A. W. S., Chan, C. C. H., and He, J. 2004. Structural stability and reliability of the Swedish occupational fatigue inventory among Chinese VDT workers. *Applied Ergonomics*, 35: 233–241.

Lin, M. Y., Catalano, P., and Dennerlein, J. T. 2016. A psychophysical protocol to develop ergonomic recommendations for sitting and standing workstations. *Human Factors* 58(4): 574–585.

Löffler, D., et al. 2015. Office ergonomics driven by contextual design. *Ergonomics in Design* 23(3): 31–35.

Mattila, M. 1996. Computer-aided ergonomics and safety—a challenge for integrated ergonomics. *International Journal of Industrial Ergonomics*, 17: 309–314.

McAtamney, L., and Corlett, E. N. 1993. RULA: A survey method for the investigation of work-related upper limb disorders. *Applied Ergonomics*, 24: 91–99.

Murata, A., Uetake, A., and Takasawa, Y. 2005. Evaluation of mental fatigue using feature parameter extracted from event-related potential. *International Journal of Industrial Ergonomics*, 35: 761–770.

Paquet, V., and Feathers, D. 2004. An anthropometric study of manual and powered wheelchair users. *International Journal of Industrial Ergonomics*, 33: 191–204.

Park, M. Y., Kim, J. Y., and Shin, J. H. 2000. Ergonomic design and evaluation of a new VDT workstation chair with keyboard-mouse support. *International Journal of Industrial Ergonomics*, 26: 537–548.

Porter, J. M., Case, K., Marshall, R., Gyi, D., and Oliver, R. 2004. "Beyond Jack and Jill": Designing for individuals using HADRIAN. *International Journal of Industrial Ergonomics*, 33: 249–264.

Psihogios, J. P., Sommerich, C. M., Mirka, G. A., and Moon, S. D. 2001. A field evaluation of monitor placement effects in VDT users. *Applied Ergonomics*, 32: 313–325.

Reed, M. P., and Van Roosmalen, L. 2005. A pilot study of a method for assessing the reach capability of wheelchair users for safety belt design. *Applied Ergonomics*, 36: 523–528.

Robertson, M. M., Ciriello, V. M., and Garabet, A. M. 2013. Office ergonomics training and a sit-stand workstation: Effects on musculoskeletal and visual symptoms and performance of office workers. *Applied Ergonomics*, 44(1): 73–85.

Sengupta, A. K., and Das, B. 1997. Human: An AutoCAD based three-dimensional anthropometric human model for workstation design. *International Journal of Industrial Ergonomics*, 19: 345–352.

Sundin, A., Christmanson, M., and Ortengren, R. 2000a. Use of a computer manikin in participatory design of assembly workstations. In K. Landau (ed.) *Ergonomic Software Tools in Product and Workplace Design*. Stuttgart: Ergon Verlag, 204–213.

Sundin, A., Ortengren, R., and Sjoberg, H. 2000b. *Proactive Human Factors Engineering Analysis in Space Station Design Using the Computer manikin Jack*. Proceedings of SAE Conference on Digital Human Modelling DHMC 2000, Dearborn, MI.

Tayyari, F., and Smith, J. 1997. Occupational ergonomics: Principles and applications. In Edited Book *Manufacturing Systems Engineering Series (MSES)* (vol. 3). New York, NY: Springer.

Tilley, A. R. 2002. *The Measure of Man and Woman: Human Factors in Design*. New York: John Wiley & Sons, 36–37.

Wang, E. M. Y., Wang, M. J., Yeh, W. Y., Shih, Y. C., and Lin, Y. C. 1999. Development of anthropometric work environment for Taiwanese workers. *International Journal of Industrial Ergonomics*, 23: 3–8.

Wang, X., and Verriest, J. P. 1998. A geometric algorithm to predict the arm reaches posture for computer-aided ergonomic evaluation. *The Journal of Visualization and Computer Animation*, 9(1): 33–47.

Woodson, W. E., Tillman, B., and Tillman, P. 1992. *Human Factors Design Handbook*. London: McGraw-Hill.

Chapter 10

Quality assurance framework for smart drug delivery systems in pharmaceutical industries and its applications

Sara Khan, Juber Akhtar, Muhammad Farhan, Khujamshukurov Nortoji A., Badruddeen, Mohammad Irfan Khan, Mohammad Ahmad, and Zainab Fatima

10.1 INTRODUCTION

Total quality control basically refers to an operation or a procedure that attempts to generate a flawless product through a number of steps that call for a coordinated effort from the complete business to eliminate inaccuracy at each level of production. Although quality assurance personnel are primarily responsible for ensuring product quality, many other departments and disciplines within a business are also involved. It requires teamwork in order to be successful. Quality must be considered during designing a drug-related product and manufacturing process. It is affected by factors like the physical layout of the manufacturing facility, ventilation, cleanliness, and hygienic conditions (Lachman et al., 2013).

The terms quality control (QC) and quality assurance (QA) are frequently used conversely. Although the two concepts are comparable, they also differ significantly. The significant difference is as follows (Figure 10.1).

This chapter extensively examines the importance of quality assurance and quality control in ensuring the reliability of pharmaceutical products. It further delves into the development of a quality-assured framework, specifically focusing on smart drug delivery systems. The discussion encompasses various types of DDS, such as transdermal and carrier-based systems, which include liposomes, ethosomes, nanoparticles, and microspheres. Additionally, the chapter explores the advancements made in implantable drug delivery systems and nasal drug delivery systems. Finally, the practical applications of smart drug delivery systems are explored, emphasizing their potential to transform healthcare delivery significantly.

10.1.1 Quality assurance

Quality Assurance warrants that process employed and customer expectations are in close conformity. Quality assurance instils trust on dual fronts:

 DOI: 10.1201/9781003495314-10

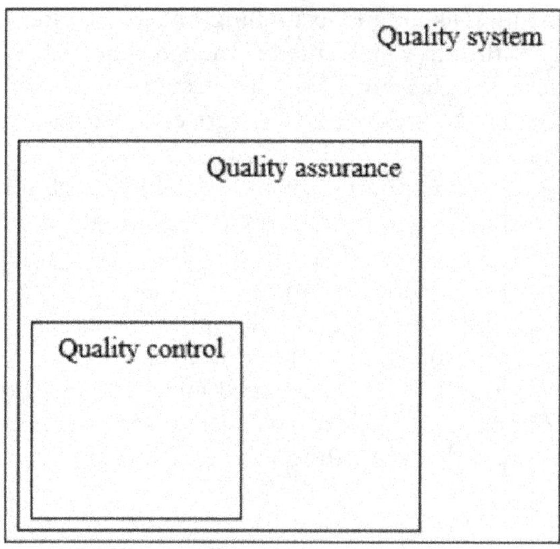

Figure 10.1 Quality System, Quality Assurance, and Quality Control Relationships.

internally, within management, and externally, with customers, regulatory bodies, and other external stakeholders (Lance et al., 2015).

10.1.2 Quality control

According to ISO 9000 definitions from ISO 9000:2015: Quality Management Systems—Fundamentals and Vocabulary quality control can be defined as "part of *quality management* focused on fulfilling *quality requirements.*" (Lance et al., 2015).

In order to guarantee the patient's safety and the medication's efficacy, the quality control of the pharmaceutical industry is crucial. Drug research, production, and distribution all depend heavily on quality control (Tailor et al., 2023).

10.1.3 Quality assurance framework

It is a set of procedures and standards created by an organization to ensure that the products or services it produces meet certain quality standards. The framework usually involves a combination of internal and external audits, inspections, and reviews of processes, procedures, and outputs to ensure that they meet or exceed established quality standards.

Some common elements of a quality assurance framework include-

Quality planning: This involves setting quality objectives, establishing procedures to meet those objectives, and determining the resources needed to implement the procedures.

Quality Control: This entails examining and evaluating goods and services to make sure they live up to predetermined standards of quality.

Quality Assurance: This involves a more comprehensive review of processes and procedures to ensure that they are effective and efficient in meeting quality standards.

Continuous improvement: This involves ongoing evaluation and improvement to ensure processes and procedures are up-to-date and effective.

Training and development: This includes providing training and development opportunities to staff members to ensure they have the skills and knowledge needed to produce high-quality products or services.

The specific elements of a quality assurance framework may vary depending on the organization's size, industry, and specific quality objectives. However, the overall goal is always to ensure that the organization produces high-quality products that meet the consumer expectations.

10.2 SMART SYSTEMS IN THE PHARMACEUTICAL INDUSTRY

Smart systems basically refer to the novel approaches employed in the delivery of drugs/products. In the present scenario, a wide variety of novel approaches/smart systems are used for the same.

Drug delivery: In layman's terms, drug delivery is a procedure of dispensing or conveying a pharmaceutical product or substance to a patient in order to accomplish its therapeutic effect in organisms—i.e., either humans or animals depending on whom it is administered.

The advantages of drug delivery systems are as follows:

- Drugs can be instilled easily.
- Bioavailability is found to be satisfactory.
- The onset of action is quick.
- Quick absorption.
- The manufacturing cost is less, and thus it is economical.
- Accurate and consistent dosing can be easily achieved.
- There is a decrease in hospital outpatient care.

The disadvantages of drug delivery systems are as follows:

- There are chances of untoward immunogenic reactions.
- In some cases, the pathology is affected adversely

(a) Conventional drug delivery system

The conventional drug delivery system is commonly referred to as the traditional drug delivery system. The conventional framework incorporates the usual approaches to drug administration, whereas the modern drug transport techniques include the following:

- Non-intrusive oral—i.e., via mouth
- Topical—i.e., via skinroutes
- Through the mucous membrane—i.e., vianose, eyes, vagina, below tongue (sublingually), rectum, and inhalation routes.

The limitations of conventional drug delivery system are as follows:

- Drug delivery must be done frequently.
- Due to the shorter half-life, there is a higher chance of missing the medication, which can decrease patient conformance.
- Reaching the steady-state condition is difficult because of the concentration-time profile of plasma.

(b) Novel drug delivery system

The novel drug delivery system refers to the non-conventional drug delivery system. The term "novel" means "newer." It is an alternative technique or new/novel strategy to the traditional drug delivery system for delivering drugs. A novel strategy for delivering pharmaceutical compounds in the body that blends creative development, formulations, new technologies, and novel techniques that are necessary to obtain their desired pharmacological effects safely is known as novel drug delivery system (NDDS). It might involve a precise scientific site for targeting inside the body, which increases the potency of pharmaceutical agents and controls the release rate of drugs.

The advantages of the NDDS over conventional drug delivery system are as follows:

- Controlled delivery and release rate of the drug by maintaining the needed concentration of the drug.
- Dosing accurately
- Site-specific drug delivery along with optimum dosing
- Reduced toxicity and side effects
- It is beneficial to patients with a good standard of living.

The various NDDSs are as follows:

- Transdermal drug delivery system
- Carrier-based delivery systems
- Implantable drug delivery system
- Nasal drug delivery system

The types of drug delivery systems are given in Figure 10.2.

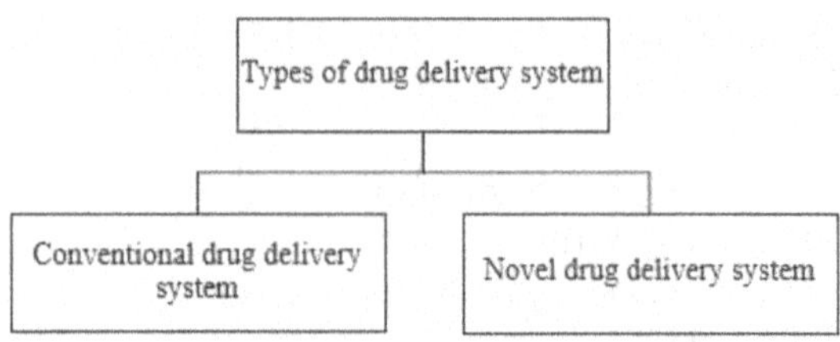

Figure 10.2 Types of Drug Delivery System.

10.2.1 Transdermal drug delivery system

A transdermal drug delivery system, also known as a transdermal patch or skin patch, is a medication delivery system that delivers a particular dose of medication to the bloodstream. It's a sticky patch that contains the medication. When drugs are delivered via the epidermis, the physicochemical properties of the skin must be taken into consideration along with the systemic effects (Patel et al., 2011). Transdermal drug delivery products provide patients with therapeutic benefits. A statistical analysis discovered a market of \$12.7 billion in 2005 and a market of \$21.5 billion in 2015, which is anticipated to increase to \$31.5 billion in 2015 (Saroha et al., 2011).

The advantages of a transdermal drug delivery system are as follows:

- Transdermal delivery of drug avoids the acidic environment of the stomach where it can undergo degradation, making it unpleasant for the patient due to various gastrointestinal symptoms or become ineffective (Gordon et al., 2005).
- Transdermal delivery prevents the degradation or metabolism of the drug before reaching systemic circulation (Rios et al., 2007).
- Transdermal delivery of medication maintains consistent plasma levels (Gordon et al., 2005).

The disadvantages of a transdermal drug delivery system are as follows:

- Transdermal patches instilled with metals have a tendency to burn skin during shock therapy carried out by defibrillators (Patel et al., 2012).
- As per the current scenario, only small molecules with affinity for lipids can pass through the layers in the body using patches. With advancements in technology, drug molecules are becoming complex; thus, new methods are required for their delivery through the skin (Prausnitz et al., 2008).

10.2.2 Carrier-based delivery systems

Wide advancements in drug delivery systems in recent years have allowed simpler routes of administration. Drug carriers are used to carry or transport the medicine to tissues that are specific for it (Suresh et al., 2013).

Various drug delivery carriers are as follows:

- Liposomes
- Ethosomes
- Nanoparticles
- Microspheres
- Polymeric micelle formulations

10.2.2.1 Liposomes

Liposomes are spherical, perishable compartments with sizes ranging from low micrometers to tens of micrometers. They are accompanied by a bilayer membrane that entraps a binary compound center. Liposome membranes are sometimes made from phosphatides, or from pure synthetic lipids with defined chemical group chains and head teams. The medication can be contained at the junction of two layers, in the volume of a binary molecule, or in the bilayer of phospholipids. Phospholipid-based liposomes are frequently employed to change the properties of drugs, enzymes, and other compounds. By raising drug concentration levels in growth cells, it is very helpful in enhancing the therapeutic effect of anti-cancer drugs. Liposomes are essential for increasing the solubility, bioavailability, and duration of drug release (Chen, 2012).

Lipids or fat molecules enclosed in a water core make up liposomes. These liposomes are employed in the treatment of infectious diseases, the manufacture of vaccines, and cancer therapy.

Liposomes have the disadvantage of leakage, which results in poorly controlled release and less encapsulation capacity (Liu, 2012).

Various liposomal drug delivery systems with liposome formulations and names of drugs loaded with targets and activities have been given in Table 10.1.

10.2.2.2 Ethosomes

Ethosomes are a very minor modification of the drug-delivery Liposome. Ethosomes are phospholipid carriers with a high ethanol concentration (20–45%) that increase drug permeability through the skin by liquefying skin lipids10. Ethosomes are essentially a mixture of ethanol, phospholipids, and water that help drugs penetrate the epidermis more effectively. It causes blood circulation to improve and drug delivery to reach deeper layers of the skin. The dimension range of ethosomes can range from nanometers to microns. According to the requirements of the patient, ethosomes can administer medications in the form of gel or cream.

10.2.2.3 Nanoparticles

Nanoparticles are sub-micron-sized particles with sizes ranging from 10 to 200 nm that exist in solid form. Because of their stability and potential

Table 10.1 Liposomal Drug Delivery System.

S. No	*Liposomal formulation*	*Drug loaded*	*Target*	*Activity*	*References*
1	Stearyl amine and diacetyl phosphate	Zidovudine	HIV	Targeting to ZDV lymphatics is enhanced	(Takemoto et al., 2004)
2	Hydrogenated soy phosphatidylcholine cholesterol and distearoylphosphatidyl glycerol	Amphotericin B	*Aspergillus fumigates*	Targeted delivery of drug at the infection site	(Cabanes et al., 1998)
3	Partially hydrogenated egg phosphatidylcholineand cholesterol	Gentamycin	*Klebsiella pneumonia*	Drug showed an increased survival rate of animal model and increased therapeutic efficacy	(Omri et al., 2002)

storage, nanoparticles are widely used as carriers. They will encapsulate or absorb the medication, shielding it from chemical and catalyst breakdown or decomposition. Nanoparticles are made up of nanocapsules and nanospheres. The nanocapsules are sac-like structures enclosing the drug within a chemical membrane, whereas the drug spreads out evenly within the nanosphere matrix systems. Nanoparticles are extremely cost-effective in transporting both hygroscopic and water-repellent medications (Shih et al., 2011).

10.2.2.4 Microspheres

These are endogenous and intra-arterial targeted drug delivery devices made from solid chemical compound matrices (Table 10.2). Microspheres are circular in shape and range in size from 1–300 m. Albumin, starch, gelatin, dextran, polypropene, and other polysaccharides are commonly used in the production of microspheres. The degradation and dissolution of the matrix regulate medication administration via these microspheres. The polymer variety and matrix size have an impact on drug delivery (Barakat et al., 2012).

10.2.2.5 Polymeric micelle formulations

The polymeric micelle has excellent characteristics in terms of delivering drugs and is easily handled. They are excellent drug carriers with an interior center lacking affinity for water and an outer corona with an affinity for water. The inner center can dissolve lipid-soluble substances and is stabilized by chemical compound chains that face the aqueous environment and thus have an affinity for water. The outer layer connects the inner core with its surroundings in the binary complex (Eshita et al., 2011).

10.2.3 Implantable drug delivery systems

A medical implantable device placed within a patient's tissues to administer therapeutic substances, improving their effectiveness and safety by regulating the timing, rate, and site of drug release in the body is known as an implantable drug delivery system (Jain, 2008).

IDDS connects the biomolecular target and the medication repository. Importantly, IDDSs combine two or more regulated components to operate as a single entity for regulatory purposes (Choi et al., 2018).

Table 10.2 List of Drugs that Are Given as Microspheres.

S. No	*Drug*	*Category*	*Polymer*	*References*
1	Metformin HCL	Antidiabetic	Sodium alginate	(Yelanki et al., 2010)
2	Amoxicillin trihydrate	Antibiotic	Ethylcellulose	(Khazeli et al., 2008)
3	Ibuprofen	Analgesic	Sodium alginate	(Kalyankar et al., 2010)

The advantages of implantable drug delivery system are as follows:

- It can easily bypass first pass metabolism.
- It improves the stability as well as bioavailability of drugs.
- It improves patient compliance.

The disadvantages of implantable drug delivery system are as follows:

- Surgery is required for the removal of large implants.
- There is a possibility of a reaction between the host and the implant.

10.2.4 Nasal drug delivery system

An intranasal drug delivery system is a form of drug delivery system that administers drugs via the nasal route. The nasal route is appealing for drug delivery because it provides numerous benefits. Intranasal drug delivery systems can be used for a wide range of medications, including those used to address migraines, pain, allergies, and respiratory diseases. These medications can be administered via intranasal drug delivery methods. Intranasal drug delivery enables the drug to reach the brain directly, avoiding profound hepatic and intestinal biotransformation. This route has been handy and dependable. Several novel formulations are being developed to transport drugs to the brain via other pathways. For example, Isotropic nanoemulsions are delivered via the intranasal route (Mehta et al., 2008).

The advantages of nasal drug delivery system are as follows:

- There is no decomposition of the drug generally observed in stomach-acidicenvironments.
- This type of delivery system provides a quick onset of action along with rapid absorption of the drug or medication.
- It is basically a non-invasive route of drug delivery.

The disadvantages of nasal drug delivery system are as follows:

- Administration of drugs by this route is relatively inconvenient to patients when compared with administration by oral route.
- When a drug is administered by this route there is a significant risk of irreversible damage to cilia on the nasal mucosa.

10.3 APPLICATION OF SMART SYSTEMS IN THE PHARMACEUTICAL INDUSTRY

A wide range of applications are as follows:—

10.3.1 Application of phytosomal formulation

- Increased absorption. E.g.-Ginseng phytosome
- Increased bioavailability. E.g.-Curcumin phytosome (Thapa et al., 2013)

10.3.2 Application ofethosome formulations

- Improved drug delivery facilitates deeper drug infiltration into the skin. E.g.-*Sophora alopocyroidsare* used as an anticancer agent
- Enhanced anti-inflammatory activity. E.g., *Glycirrhiza glabra* (Bhokare et al., 2016)

10.3.3 Application of microsphere formulation

- Higher bioavailability. E.g.-Zeodary oil microsphere
- Targeting into the cardiovascular region. E.g.-Rutin—alginate chitosan (Dilpreet, 2015)

10.3.4 Application of nanoparticles

- Improving cerebral blood flow and metabolism. E.g.-*Gingko biloba*
- Increased antioxidant activity. E.g.-Quercetin
- Decreasing the toxicity. E.g.-Triptolide
- Sustained drug release. E.g.-Berberine (Ajazuddin, 2010)

10.3.5 Application of nanoemulsions

- Drug bioavailability is increased by nano-emulsions of antitubercular medications, which can readily penetrate biological barriers to reach systemic Mycobacterium tuberculosis infection (Beg et al., 2017).
- A potential method for increasing vitamin D bioavailability is the nano-emulsion delivery system (Kadappan et al., 2018).
- Sustained release formulation. E.g. Silybin nanoemulsions (Chaudhary et al., 2020).
- Improved absorption. E.g. Berberine nanoemulsions (Chaudhary et al., 2020).

10.4 CONCLUSION

This chapter emphasizes the significance of quality assurance and quality control in ensuring pharmaceutical product reliability. Our main focus was on smart drug delivery systems (DDS) and the exploration of various types, including transdermal and carrier-based drug delivery systems like liposomes, ethosomes, nanoparticles, and microspheres, underscoring the need for maintaining high standards in the pharmaceutical industry. The advancements in implantable drug delivery systems (IDDS) and nasal drug delivery systems (NDDS) hold promise for future drug delivery technologies. Overall, the practical applications of smart drug delivery systems have the potential to transform healthcare delivery by improving treatment efficacy, minimizing side effects, and ultimately enhancing patient outcomes on a personalized level.

10.5 FUTURE PERSPECTIVE OF THE QUALITY ASSURANCE FRAMEWORK IN SMART SYSTEMS

Pharmaceutical manufacturers bear enormous professional and legal responsibilities for ensuring product excellence. The utmost requirement to ensure a progressive future of any operation depends on ensuring a regimented management system. Only through professional, disciplined staff, well-ordered process, and methodical control of dosage form, before, during, and after production, a requisite product quality assurance can be obtained. Product quality cannot be solely determined by testing of dosage form and its control unless a definite process control is efficiently applied; thus, good manufacturing practices are a prerequisite for it.

The pharmaceutical manufacturer has the chief authority in terms of the quality of goods produced under his effective guidance. Through a well-organized total quality assurance system, the manufacturer is able to (1) administer the root cause of variation in the quality of the product like materials used, methods applied, and manpower involved, (2) ensure pertinent manufacturing and packaging practices, (3) ensure that the results of various tests performed are in accordance with the official standards, and (4) ensure the stability of the product.

Certain basic operational rules must be established and followed in order for the total quality assurance system to operate properly. First, all the decisions regarding the control must be based on the quality of the product and other considerations related to it. Second, the operation must be stringently carried out as per the set standards, and it must always endeavor to improve current standards. Third, sufficient provisions, personnel capitalization, and an environment for personnel to accomplish their responsibilities should be provided. Last but not least, control decisions must never be revoked by production or marketing employees. The climate required for making sound decisions is critical because the control choice can affect both the consumer's health and the manufacturer's reputation. Control choices should only be reviewed at the highest level of management in times of major disagreement.

The act of seeking study on NDDS has been in place for a long time, but it has only recently gained power. Several kinds of NDDS have been developed in recent decades, including liposomes, nanoparticles, microparticles, niosomes, and microencapsulation, with the aim of conveying drugs to specific tissues in order to cure diseases. Numerous research opportunities in nanotechnology and related fields have been made possible by recent advancements in the field of NDDS. It is important to note that the effective treatment of cancer in its various forms by targeting the pretentiously affected cells is one of the most promising uses of NDDS.

ACKNOWLEDGMENT

The author thanks Integral University, India, for providing facilities and support in writing this chapter. They are grateful to Professor (Dr,) Juber Akhtar of Integral University, for his guidance, advice, and fruitful discussion throughout the chapter.

REFERENCES

Ajazuddin SS. Application of novel drug delivery system for herbal formulation. *Fioterpia*, 2010; 81(7): 680–689. DOI: https://10.1016/j.fitote.2010.05.001;PMID:20471457

Barakat NS, et al. Target nanoparticles: an appealing drug delivery platform. *Journal of Nanomedcine & Nanotechnology*, 2012; S4:009. https://www.rroij.com/open-access/new-trends-drug-delivery-systems-.php?aid=79698#83

Beg S, Saini S, Imam SS. Nanoemulsion for the effective treatment and management of anti-tubercular drug therapy. *Recent Patents on Anti infect Drug Discovery*, 2017; 12(2): 85–94.

Bhokare SG, Dongaonkar CC, Lahane SV, Salunke PS, Sawale VS, Thombare MS. Herbal novel drug delivery-a review. *World Journal of Pharmacy and Pharmaceutical Science*, 2016; 5(8): 593–611. DOI: https://10.20959/wjpps20168-7461.

Cabanes A, Reig F, Garcia-Anton JM, Arboix M. Evaluation of free and liposome-encapsulated gentamycin for intramuscular sustained release in rabbits. *Research in Veterinary Science*, 1998; 64: 213–217.

Chaudhari PM, Randive SR. Incorporated herbal drugs in novel drug delivery system. *Asian Journal of Pharmacy and Pharmacology*, 2020; 6(2): 108–118.

Chen G. Nanotube-based controlled drug delivery. *Pharmaceut Anal Acta*, 2012; 3: e136. https://www.rroij.com/open-access/new-trends-drug-delivery-systems-.php?aid=79698#83

Choi SH, Wang Y, Conti, DS, Raney SG, Delvadia R, Leboeuf AA, Witzmann K. Generic drug-device combination products: Regulatory and scientific considerations. *International Journal of Pharmaceutics*, 2018. https://www.mdpi.com/23065354/8/12/205#:~:text=The%20implantable%20drug%20delivery%20system,in%20the%20body%20%5B4%5D.

Dilpreet S. Application of novel drug delivery system in enhancing the therapeutic potential of phytoconstituents. *Asian Journal of Pharmaceutics*, 2015; 9(4): $1–$12.

Eshita Y, et al. Mechanism of the introduction of exogenous genes into cultured cells using DEAE-Dextran-MMA graft copolymer as a non-viral gene carrier. II. It's thixotropy property. *Journal of NanomedicNanotechnol*, 2011; 2: 105. https://www.rroij.com/open-access/new-trends-drug-delivery-systems-.php?aid=79698#83

Gordon RD. More than skin deep: advances in transdermal technologies are opening up new avenues of exploration. *Pharmaceutical Technology Europe*, 2005; 17(11): 60–66.

Jain, K.K. Drug delivery systems—an overview. *Methods Molecular Biology*, 2008; 437: 1–50. https://www.mdpi.com/23065354/8/12/205#:~:text=The%20implantable%20drug%20delivery%20system,in%20the%20body%20%5B4%5D.

Kadappan AS, Guo C. The efficacy of Nano emulsion-based delivery to improve vitamin D adsorption: comparison of in vitro and in vivo studies. *Molecular Nutrition, and Food Research*, 2018; 62(4).

Kalyankar TM, Nalanda T, Mubeena K, Hosmani A, Sonawane A. Formulation and evaluation of mucoadhesive pioglitazone Hcl microspheres. *International Journal of Pharma World Research*, 2010; 1(3): 1–14.

Khazaeli P, Pardakhty A, Hassanzadeh F. Formulation of Ibuprofen beads by ionotropic gelation. *Iranian Journal of Pharmaceutical Research*, 2008; 7(3): 163–170.

Lachman L, Lieberman HA, Kaing JL. *The theory and practice of industrial pharmacy*. New Delhi CBS Publishers and Distributors Pvt. Ltd, 2013.

Lance B. Coleman Sr. *ASQ/ANSI/ISO 9000:2015: Quality management systems—fundamentals and vocabulary*, 2015. https://asq.org/quality-resources/quality-assurance-vs-control

Liu R. Nanostructured lipid carriers as the most promising approach in ocular drug delivery system. *Journal of NanomedBiotherapeutic Discovery*, 2012; 2: e116. https://www.rroij.com/open-access/new-trends-drug-delivery-systems-.php?aid=79698#83

Mehta SK, Kaur G, Bhasin KK Incorporation of antitubercular drug isoniazid in pharmaceutically accepted microemulsion: Effect on microstructure and physical parameters. *Pharmaceutical Research*, 2008; 25: 227–236. DOI: https://10.1007/s11095-007-9355-8

Omri A, Suntres ZE, Shek PN. Enhanced activity of liposomal polymyxin B against Pseudomonas aeruginosa in a rat model of lung infection. *Biochemical Pharmacology*, 2002; 64: 1407–1413.

Patel D, Chaudhary SA, Parmar B, Bhura N. Transdermal drug delivery system: a review. *The Pharma Innovation*, 2012; 4(1): 78–87.

Patel DM, Kavitha K. Formulation and evaluation aspects of transdermal drug delivery system. *International Journal of Pharmaceutical Sciences Review and Research*, 2011; 6(2): 83–90.

Prausnitz MR, Langer R. Transdermal drug delivery. *Nature Biotechnology*, 2008; 26(11): 1261–1268.

Rios M. Advances in transdermal technologies: transdermal delivery takes up once-forbidden compounds, reviving markets and creating formulation opportunities. *Pharmaceutical Technology*, 2007; 31(10): 54–58.

Saroha K, Yadav B, Sharma B. Transdermal patch: a discrete dosage form. *International Journal of Current Pharmaceutical Research*, 2011; 3(3): 98–108.

Shih MF, et al. Bioeffects of transient and low-intensity ultrasound on nanoparticles for a safe and efficient DNA delivery. *Journal of Nanomedicine Nanotechnology*, 2011; S3: 001. https://www.rroij.com/open-access/new-trends-drug-delivery-systems-.php?aid=79698#83

Suresh Kumar R, et al. Self-nano emulsifying Drug delivery system of olanzapine for enhanced oral bioavailability: in vitro, in vivo characterisation and in vitro -in vivo correlation. *Journal of Bioequivalence*, 2013; 5: 201–208. https://www.rroij.com/open-access/new-trends-drug-delivery-systems-.php?aid=79698#83

Tailor P, Parmar A. Exploring the evolution of quality control in pharmaceuticals: What has changed? *Research & Review: Drugs and Drugs Development (e-ISSN: 2582–5720)*, 2023; 23–27.

Takemoto SK, Zeevi A, Feng S, Colvin RB, Jordan S, et al. National conference to assess antibody-mediated rejection in solid organ transplantation. *American Journal of Transplantation*, 2004; 4: 1033–1041.

Thapa RK, Khan GM, Baralk P, Thapa P. Herbal medicine incorporated nanoparticles: advancements in herbal treatment. *Asian Journal of Biomedical and Pharmaceutical Sciences*, 2013; 3(24): 7–14.

Yellanki SK, Singh J, Syed JA, Bigala R, Garanti S. Design and characterization of Amoxicillin trihydrate mucoadhesive microspheres. *International Journal of Pharmaceutical Science and Drug Research*, 2010; 2(6): 112–114.

Chapter 11

Multi-agent-based dynamic scheduling of flexible manufacturing systems with routing flexibility

Dr. Mohd. Shaaban Hussain and Dr. Mohammed Ali

11.1 INTRODUCTION

Industry 4.0, also known as I4.0, pertains to the digitization of manufacturing and comparable businesses aimed at optimizing value processes through an increasing focus on technology such as test automation services and data interchange. The manufacturing systems are now more versatile due to the adoption of Industry 4.0 technology. The Internet of Things (IoT), big data, artificial intelligence (AI), additive manufacturing (AM), sophisticated robots, virtual reality, cloud computing, and simulation are a few examples of important technology. Industry 4.0 is now universally acknowledged to be a distinctive industrial paradigm built on the broad use of communication and information technologies, which in turn improves organizational performance and adaptability. In terms of boosting flexibility and customization within the industrial system, the introduction of Industry 4.0 technologies is particularly transformative (Fragapane et al. 2022; Khan and Ali 2015).

The domain that has experienced the most significant influence from I4.0 is manufacturing. Presently, the manufacturing industry is undergoing a transformation away from the traditional approach of centralization and mass production, which was the cornerstone of the first three industrial revolutions, to a more customer-centric approach focused on mass customization. This new approach entails producing goods as close as possible to the areas of demand (Groover 2016; Li et al. 2021).

Industry 4.0 is predominantly based on cyber-physical systems (CPS) technology, which enables modular and adaptable production systems capable of mass-producing highly customized products (Kagermann 2014; Nascimento et al. 2019). CPS makes it possible for the physical and virtual worlds to merge by utilizing the IoT to connect infrastructure, real-world items, human actors, machines, and processes across organizational boundaries. With the help of sensors, actuators, and computing power, this integration transmits real-time data for distributed decision-making processes (Trappey et al. 2017). CPS, which includes the decision-making process in production contexts, gives one the ability to respond autonomously and flexibly to

DOI: 10.1201/9781003495314-11

unforeseen circumstances. In addition, CPS creates virtual versions of the manufacturing process, enabling remote management via cloud computing (Wang et al. 2014; Gao et al. 2015), a crucial aspect of cloud manufacturing (Xu 2012; Wang and Wang 2014; Wang et al. 2015; Liu et al. 2019). The concept of cyber-physical production systems (CPPS) is the result of these and other CPS applications (Lee et al. 2015).

The development of CPPS can be attributed to the use of CPSs in manufacturing. Monostori (2014) asserts that, depending on the context, CPPS consists of autonomous and cooperative elements and subsystems that are integrated to cover all phases of the manufacturing process, from the shop-floor to the logistics networks. In the context of Industry 4.0, the CPPS stands for a collection of interconnected subsystems. The capacity of CPPS to connect the shop floor directly with a higher-level decision support system (DSS) is one of its main advantages (Rossit and Tohmé 2018). The shop floor can swiftly adjust to the DSS's output thanks to this link, which gives the DSS real-time data.

Industry 4.0 is distinct from its predecessors in that it utilizes decentralized architectures with autonomous agents that interact with decision centers, as opposed to traditional hierarchical and centralized structures. In software engineering and automation, multi-agent systems (MASs), a modeling framework suited for distributed systems, have been used. MAS achieves modularity, autonomy, flexibility, resilience, and adaptability by relying on agents with separate decision-making and computation skills to work together to fulfil system tasks (Hussain and Ali 2019a; Qin et al. 2016). In terms of functionality and operation mechanism, the MAS provides a distributed system architecture design methodology that is in line with the features and specifications of CPPS. The MAS is a useful instrument for investigating production process control, quality monitoring, adaptability, reconfigurability, and self-healing capability since it is a crucial enabling technology for CPPS (Leitão et al. 2015).

11.2 INDUSTRY 4.0 AND FLEXIBLE MANUFACTURING SYSTEMS

To effectively respond to new market environments, companies need to implement digital technologies and undergo a digital transformation. This will not only improve their competitiveness but also strengthen their ability to make optimal decisions. Industry 4.0 has been adopted in the manufacturing sector with the goal of advancing industrial growth and boosting corporate competitiveness. At higher levels of organization, such as computer integrated manufacturing (CIM), CPPS can be integrated with flexible manufacturing systems (FMS), allowing for the self-regulation and improvement of production processes. These systems' adaptability and flexibility make it possible for them to swiftly adjust to a setting where changes are constant, changeable, and unpredictable. By efficiently reacting to market demands, FMS is regarded as the only manufacturing system that has found a way to reconcile the tension between productivity, flexibility, and adaptability. Important contributions have recently been made to the development of flexible manufacturing.

Simulation is a powerful tool in the context of digitalization in manufacturing. It is now feasible to simulate production systems to learn how they will react to different factors and unforeseen circumstances. When developing an analytical solution to a problem is impossible or extremely challenging, simulation models are frequently utilized. Simulation is a suitable approach to address challenging issues that develop in these systems because of the complex and dynamic behavior of FMS and the dearth of analytical models for their design, analysis, and optimization. Modeling and simulating systems with discrete events, such as FMS, have been the subject of several research. By simulating different phenomena, processes, and complex systems, researchers can gain valuable insights into these systems and improve their understanding of them.

11.3 FMS SCHEDULING PROBLEM

The FMS is comprised of six flexible machines designated as M1, M2, M3, M4, M5, and M6 (Ali and Wadhwa 2010; Hussain and Ali 2019b). Every machine has an input buffer that has a finite size. The system takes into account six different part types: P1 to P6. Every part requires four to six operations with deterministic processing times, for complete processing. For the sake of validation, we have taken the part data or part processing times from the literature (Ali and Wadhwa 2010). A product mix of 600 parts is studied, with 100 parts of each of the six part types. The only internal disturbance considered in the present problem is machine failure. The system factors considered to explore the FMS scheduling problem are as follows:

1. Routing Flexibility (RF): It measures the number of potential machines available to process a part in the system (Wadhwa et al. 2010; Chan et al. 2008; Khan and Ali 2019; Mashhood and Ali 2022). A job shop that lacks routing options is defined as having an RF of 0, whereas a job shop that has alternative routing options is defined as having an RF of at least 1. To develop a simulation model of RF and integrate it into the FMS simulation model, the concept of routing flexibility (Wadhwa and Rao 2002) is taken into consideration. Figure 11.1 illustrates the concept of routing flexibility and its levels.
2. Buffer Capacity (BC): We model buffer capacity based on the capacity of the dedicated input buffer. Four buffer capacity levels are examined, with variations of 6 steps each, ranging from 6 to 24, as determined by the input buffer size.

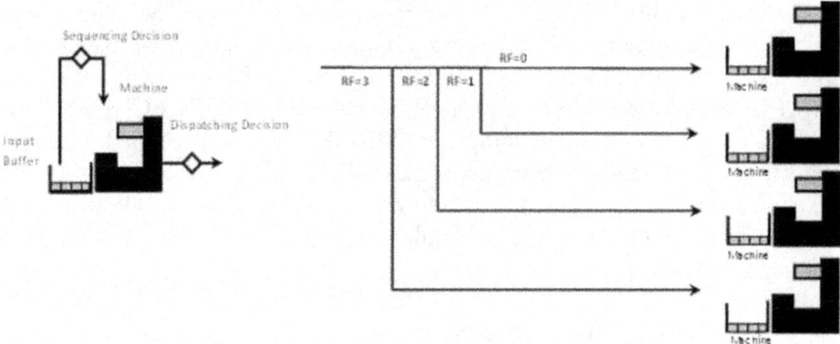

Figure 11.1 Illustration of the Concept of Routing Flexibility and Its levels.

3. Sequencing Rules (SR): The implementation of sequencing rules is crucial for efficient manufacturing control. SR plays a crucial role in determining which part from the queue is selected for processing on a machine. The following SR have been studied in the present work.
   a. First Come First Served (FCFS): A part entering the input buffer first will be processed first.
   b. Last Come First Served (LCFS): A part entering the input buffer at last will be processed at last.
   c. Shortest Processing Time (SPT): A part that requires minimum time for operation completion will be processed first from the parts available in the input buffer. This calculation is made for all parts in the dedicated input buffer of each machine.
   d. Highest Processing Time (HPT): A part that requires maximum time for operation completion will be processed first from the parts available in the input buffer. This calculation is made for all parts in the dedicated input buffer of each machine.
4. Dispatching Rule (DR): The routing of a part for its remaining processing to a potential machine is determined by the dispatching rule (Ahmad et al. 2021). The level of routing flexibility influences the dispatching decisions made. These decisions are taken once the part processing at a machine completes. In this study, the dispatching rule used is MINQ, which prioritizes the buffer with the fewest number of parts in the queue. Under this rule, the part is routed to a machine input buffer queue that has the least number of parts.

   Performance Measures: The performance of the FMS is evaluated using the following metrics:
   a. Makespan Time (MST): It is the time required to complete processing of a batch of jobs. It is the difference of time at which the last part exits the manufacturing system and the time at which the first part enters the manufacturing system.
   b. Machine Utilization (MU): It is the average utilization of the all the machines present in the layout. It is the period of time when machine may be used for processing rather than remaining idle.

In order to evaluate the performance of the system, simulation experiments have been conducted, with three system factors, each having four levels. The four levels of routing flexibility (RF0, RF1, RF2, RF3), four buffer capacity levels (6, 12, 18, 24), and four part-sequencing rules (FCFS, LCFS, SPT, and HPT), along with one dispatching rule (MINQ), have been studied. Machine utilization (MU) and makespan time (MST) have been used to gauge the system's performance. The discrete-event-based simulation models for both conventional and multi-agent scheduling approaches are based on the following assumptions:

- The processing times are deterministic.
- All machines are unreliable.
- Machine failure and repair follow an exponential distribution.
- Pre-emption is allowed.
- Only one operation can be performed on a machine at a time.
- The transfer time for each part in the system is unity.
- No consideration of due dates.
- Order cancellation is not allowed.
- Rework is not considered.

11.4 FRAMEWORK FOR SIMULATION MODEL

In this study, discrete-event simulation models based on conventional scheduling and multi-agent scheduling were developed in MATLAB. These models examine the impact of different system factors on the performance of the FMS. In the simulation models, parts, machines, and their related properties are modeled as entities. The simulation models capture the system dynamics through the generation of various events. A comprehensive validation process has been carried out to verify the accuracy of the simulation models. In multi-agent scheduling model machines are modelled as agents. These machine agents are designed to be autonomous, intelligent, and cooperative. They coordinate and exchange information among themselves to make decisions and achieve their goals. In case of disturbances, these machine agents make decisions independently, thus providing real-time control.

There are various modules in the simulation model, each with a specific task. The simulation clock is initiated by the initialization module which sets the initial values for system variables and counters. The main program invokes the control architecture module after initialization, outlines the decision-making entities' relationships and defines the control architecture. The event function schedules events such as part processing, machine breakdown, and part routing during simulation execution. The rules and factors, such as routing flexibility, are specific to each part and are implemented by the main program. After each event, the simulation clock and system state are updated and all statistical counters and variables are updated. The simulation model continually checks for termination criteria, ending the program and generating a performance report when criteria are met. If not, the simulation continues until termination criteria are satisfied.

11.5 RESULTS AND DISCUSSION

In the present study, two scheduling approaches, namely conventional and multi-agent based scheduling, are considered. The two scheduling approaches are compared and multi-objective optimization has been done using desirability approach (DA). The system factors considered for the study are routing flexibility (RF), buffer capacity (BC), and sequencing rules (SR), while makespan time (MST) and machine utilization (MU) are the performance measures. The dispatching rule used throughout the study is minimum number in queue (MINQ).

11.5.1 Design of experiment

For conducting simulation experiments, Taguchi's technique for experiment design was followed. Matrix experiments serve as the foundation for Taguchi's framework for the design of experiments (Phadke 1989). The Taguchi's matrix experiments provide us with a set of system variables at which the best system performance is possible. Using this method, a series of tests are carried out by changing the levels of the assumed variables. Matrix experiments are a useful method for determining the impact of different elements. The results of the matrix experiments are used to perform ANOM (analysis of means) and ANOVA (analysis of variance) in Taguchi's method. By statistically analyzing the findings using an ANOVA, the factors influencing the system performance are discovered. The normality of data distribution and the equal variances are checked prior to ANOVA statistical analysis.

The Taguchi's L16 orthogonal array, as illustrated in Table 11.1, was used in the simulation trials. Thirty simulation replications have been shown to be adequate to remove the random influences from the outcomes. So, for each of the 16 simulation experiments, a simulation run consists of 30 replications. To determine the importance of the current research, statistical analysis is used to the simulation data. MINITAB is used to do the computations for the ANOM and ANOVA analyses at a 95% confidence level (i.e., a significance level (α) = 0.05). Table 11.2 presents the simulation results for the conventional and multi-agent based scheduling approaches with machine failure.

As evident from Table 11.2, for all the matrix experiments conducted, the MST and MU values for multi-agent based scheduling approach are significantly better than the conventional scheduling approach. This is because under a traditional scheduling system, decision-making entities are not autonomous, and delays arise while receiving and putting orders from

Table 11.1 Taguchi Matrix Experiments.

Experiment No	*RF*	*BC*	*SR*
1	0	6	FCFS
2	1	12	LCFS
3	2	18	SPT
4	3	24	HPT
5	0	12	SPT
6	1	6	HPT
7	2	24	FCFS
8	3	18	LCFS
9	0	18	HPT
10	1	24	SPT
11	2	6	LCFS
12	3	12	FCFS
13	0	24	LCFS
14	1	18	FCFS
15	2	12	HPT
16	3	6	SPT

Table 11.2 Simulation Experiment Results.

	Conventional scheduling		*Multi-agent scheduling*	
Experiment No	*MST (min)*	*MU (%)*	*MST (min)*	*MU (%)*
1	53809.32	42.50	38831.63	59.74
2	50937.07	45.06	39422.34	58.80
3	66246.23	34.02	42098.03	55.18
4	52467.73	43.67	38796.73	60.00
5	51500.35	44.50	29619.36	79.65
6	51839.35	44.22	28101.51	84.57
7	47693.48	48.44	30147.29	78.11
8	47107.99	49.04	29603.05	79.67
9	47546.39	48.50	27364.02	86.82
10	47244.84	48.74	28508.25	83.08
11	52028.62	44.04	28929.30	81.43
12	51644.29	44.49	27997.54	84.46
13	46968.64	48.98	27721.86	85.68
14	48820.47	47.07	28204.78	83.90
15	51789.17	44.32	28753.48	82.29
16	50847.25	45.07	29605.98	79.46

higher-up decision-making entities into practice. When real-time decision-making and implementation are prevented, this causes a delay that worsens system performance. The performance of the system has improved with the multi-agent based scheduling strategy, nevertheless (Tran et al. 2019). It may be explained by the fact that real-time decision-making and its execution can be carried out, if any disturbance (machine failure) is taking place or not, as a consequence of the delegation of decision-making to subordinate decisional entities, in our case, machine agents. This leads to improved system performance. This finding is consistent with those of other studies (Barbosa et al. 2015; Rey et al. 2014; Borangiu et al. 2015; Pach et al. 2014; Zbib et al. 2012). From this point onward, we will be focussing on the multi-agent based scheduling approach only.

11.5.2 Determination of optimum system factor levels

Finding an ideal system factor combination that provides the optimum system performance with the fewest simulation trials is the fundamental goal of matrix design framework experimentation. The term "main effect" refers to the impact of a system factor level that results in a deviation from the mean. ANOM identifies each system factor's main effect. The main effects plot for each factor level in Figure 11.2 indicates that the optimal combination of

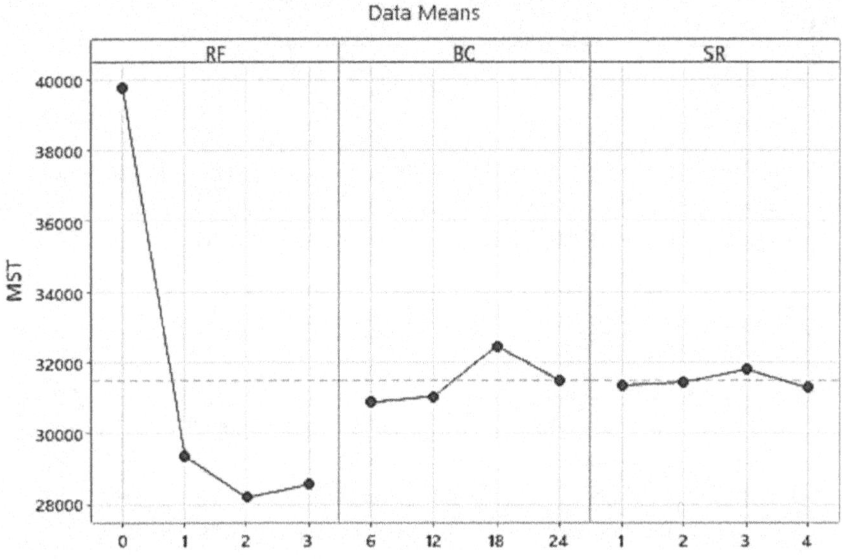

Figure 11.2 Mean Data Analysis for MST.

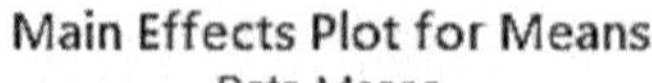

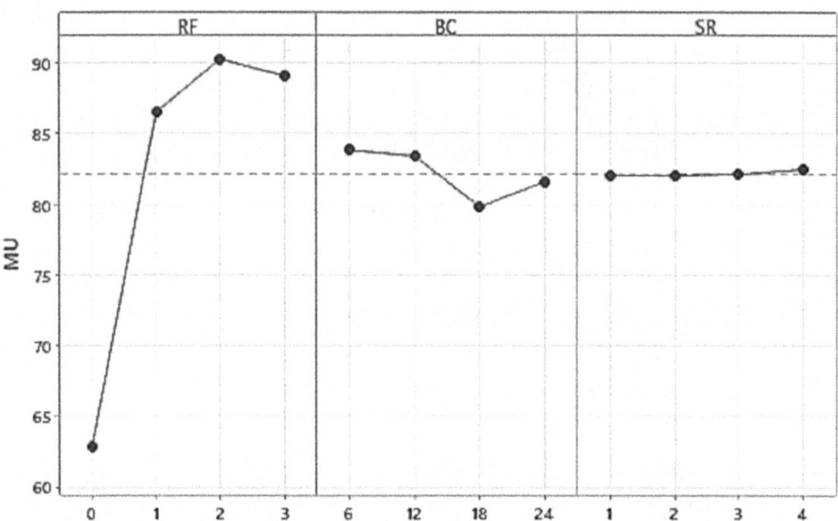

Figure 11.3 Mean Data Analysis for MU.

Table 11.3 Optimum System Factor Levels.

Response	*Routing flexibility level*	*Buffer capacity*	*Sequencing rule*
MST	2	6	HPT
MU	2	6	HPT

factors for MST is RF2, BC1, and SR4, whereas the optimal combination of factors, Figure 11.3, for the MU is RF2, BC1, and SR4. Table 11.3 summarizes the optimum system factor levels for MST and MU.

11.5.3 Determination of optimum system factor levels

The normality of data distribution and the equal variances are the two assumptions which are checked prior to ANOVA statistical analysis. The normality of the data can be confirmed with the help of normality plot of residuals. The residuals versus fitted value plot is used to confirm constant variance. A straight-line plot of residuals confirms the normality of the data. If the residuals versus fitted value plot does not follow any definite pattern, then the assumption of constant variance is confirmed.

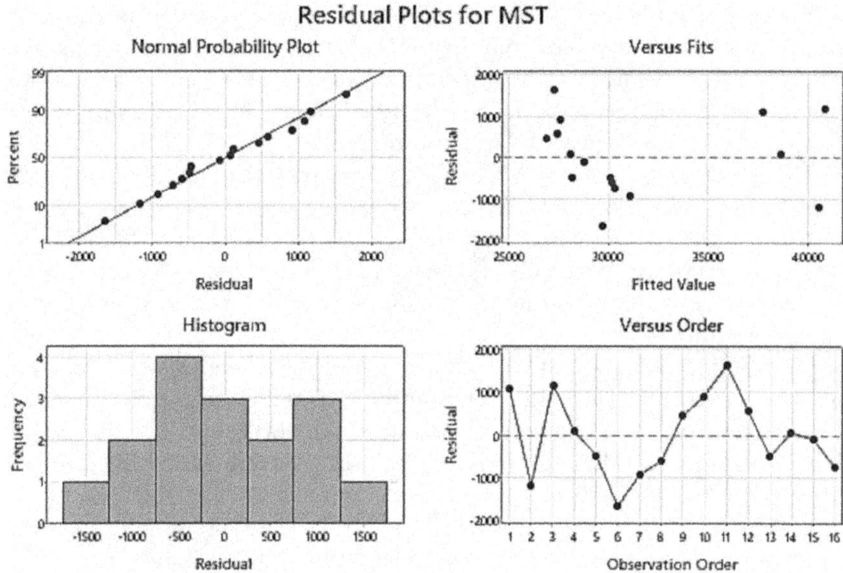

Figure 11.4 Residual Plots for MST.

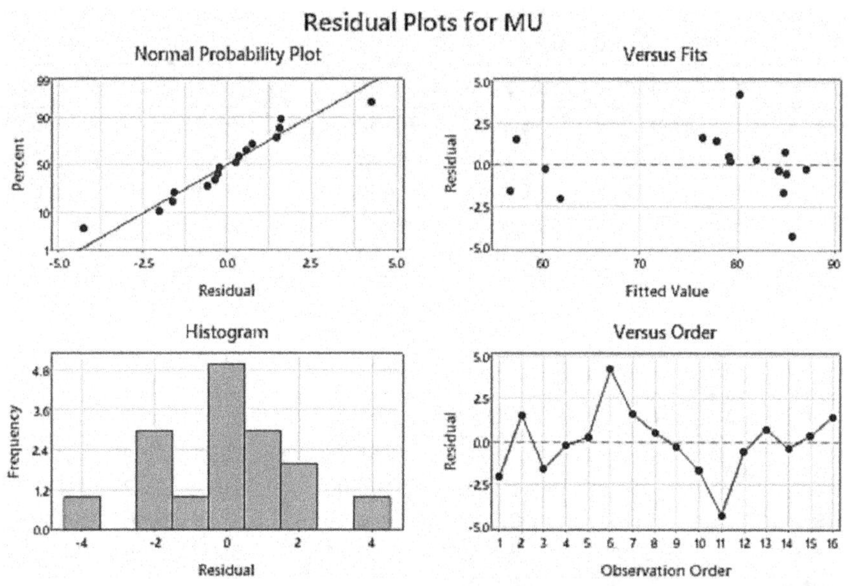

Figure 11.5 Residual Plots for MU.

The residual plots for MST and MU are shown in Figures 11.4 and 11.5, respectively. It is observed from the plots that residuals follow normal distribution. The residual versus fitted value plots for MST and AMU do not follow any definite pattern. Thus, it can be stated that the assumptions for data normality and constant variance are fulfilled.

The ANOVA analysis for MST and MU are presented in Tables 11.4 and 11.5, respectively. The simulated data from the matrix experiments are used to perform the ANOVA analysis. The F-value shows how much a factor has an impact on the performance of the system (Phadke 1989). The significance of a factor is established if its p-value is lesser than or equal to the significance level ($\alpha = 0.05$).

From Tables 11.4 and 11.5, RF has the highest F-value of 119.07 and 88.35, respectively. The system's performance is therefore greatly influenced by changes in RF level. In comparison to RF, the other two factors seem to have a less impact. The relative importance of the factors for MST and MU after the ANOVA analysis is in perfect accord with that obtained by the ANOMA, as shown in Figures 11.2 and 11.3. The most significant variations in the curve under the RF were seen in Figures 11.2 and 11.3, which suggests that the RF has a larger impact on the system than other parameters. Relative factor importance can be assessed in a similar way for other system factors.

From Tables 11.4 and 11.5, the p-value of RF is 0.00, for both for MST and MU. The p-value of RF is lesser than the significance level value of

Table 11.4 Analysis of Variance for MST.

Source	*DF*	*Seq SS*	*Contribution*	*Adj SS*	*Adj MS*	*F-Value*	*P-Value*
RF	3	370758878	96.61%	370758878	123586293	119.07	0.000
BC	3	6145979	1.60%	6145979	2048660	1.97	0.219
SR	3	649617	0.17%	649617	216539	0.21	0.887
Error	6	6227745	1.62%	6227745	1037958		
Total	15	383782219	100.00%				

Table 11.5 Analysis of Variance for MU.

Source	*DF*	*Seq SS*	*Contribution*	*Adj SS*	*Adj MS*	*F-Value*	*P-Value*
RF	3	1751.94	95.89%	1751.94	583.980	88.35	0.000
BC	3	35.00	1.92%	35.00	11.668	1.77	0.253
SR	3	0.47	0.03%	0.47	0.156	0.02	0.995
Error	6	39.66	2.17%	39.66	6.610		
Total	15	1827.07	100.00%				

0.05 while other factors have their p-values greater than the significance level. Thus, only RF has statistically significant effect on system performance among the system factors.

To verify the optimum results predicted by the Taguchi approach, confirmation experiments are conducted at the optimal factors level. The optimum factor combination is RF=2, BC=6 and SR=4, for both MST and MU. The predicted value (η_p) under optimum conditions is calculated using Equation 11.1.

$$\eta_p = \eta_m + \sum_{i=1}^{q} (\eta_i - \eta_m) \tag{11.1}$$

where η_m is the overall mean of data, η_i is the mean data at optimum levels and “q” is the number of significant input parameters that affects system performance. Table 11.6 presents the predicted optimal values and actual values of MST and MU for optimal system factor levels.

11.5.4 Multi-objective optimization using desirability approach

For multi-objective optimization Taguchi’s approach is not suitable. Hence, using Minitab software, multi-objective optimization based on the desirability approach (DA) was performed. The DA was established by Derringer and Suich (1980), and it is one of the most commonly used multi-objective (MO) techniques. In order to find the best system factor combination that produces the highest MU and lowest MST values, the MO was conducted. The maximum and minimum values of system factors, i.e. RF, BC, and SR are given in Table 11.6. Both the responses (i.e., MST and MU) were assigned equal weightage for MO. All response variables (y_i) must be normalized in order to calculate the desirability value (d_i), which ranges from 0 to 1. When the desirability of a response varies, so does the value of desirability. If d_i has a value of 0, then it indicates that the response is completely undesirable. On the other hand, if d_i has a value of 1, then it indicates that the response is completely desirable. Equations 11.2 and 11.3 below illustrate the individual desirability of response according to desire.

Table 11.6 Predicted and Experimental Values.

Response	*Optimal combination*	*Actual value*	*Predicted optimal value (η_p)*
MST	RF2BC1SR4	42121.12	42098.03
MU	RF2BC1SR4	55.21	55.18

For "higher the better"

$$di = \begin{cases} 0 & y_i \le y_{min} \\ \left[\dfrac{y_i - y_{max}}{y_{max} - y_{min}}\right]^r & y_{min} < y_i < y_{max} \\ 1 & y_i \ge y_{max} \end{cases} \tag{11.2}$$

For "lower the better"

$$di = \begin{cases} 0 & y_i \le y_{min} \\ \left[\dfrac{y_{max} - y_i}{y_{max} - y_{min}}\right]^r & y_{min} < y_i < y_{max} \\ 1 & y_i \ge y_{max} \end{cases} \tag{11.3}$$

where y_{max} and y_{min} represent the maximum and minimum value of yi. The form of d_i is determined by the value of "r." The value of "r" varies, changing the form of the desirability function. The value of "r" varies from 0.1 to 10, and a value larger than 1 emphasizes the desired outcome. Equation 11.4 is used to evaluate the total desirability (d) after achieving the desirability of each individual response, as shown below:

$$d = (d1 \times d2 \times d3 \times \ldots \ \ldots \ \ldots \ \ldots \times dk)^{1/k} \tag{11.4}$$

Table 11.7 gives the lower and upper bounds of the system factors studied in the present work. The results of the multi-objective optimization are displayed in Figure 11.6. With an ideal system factor combination of RF = 2.30, BC = 6, and SR = 1.0, the maximum MU and lowest MST at a desirability of 1.000 were found to be 93.40 and 23510, respectively.

After multi-objective optimization for maximum MU and minimum MST, confirmatory experiments were performed at the optimized system factor setting for results validation. For confirmatory experiments, RF=2 was utilized rather than 2.30 since it is the closest RF value to 2.30 that is currently available. The confirmatory values of MU and MST at the ideal system factor combination were determined to be 92.13% and 23832.09

Table 11.7 Lower and Upper Bounds of System Factors.

System factor	*Bounds*
Routing Flexibility	0≤RF≤3
Buffer Capacity	6≤BC≤24
Sequencing Rules	1≤RF≤4

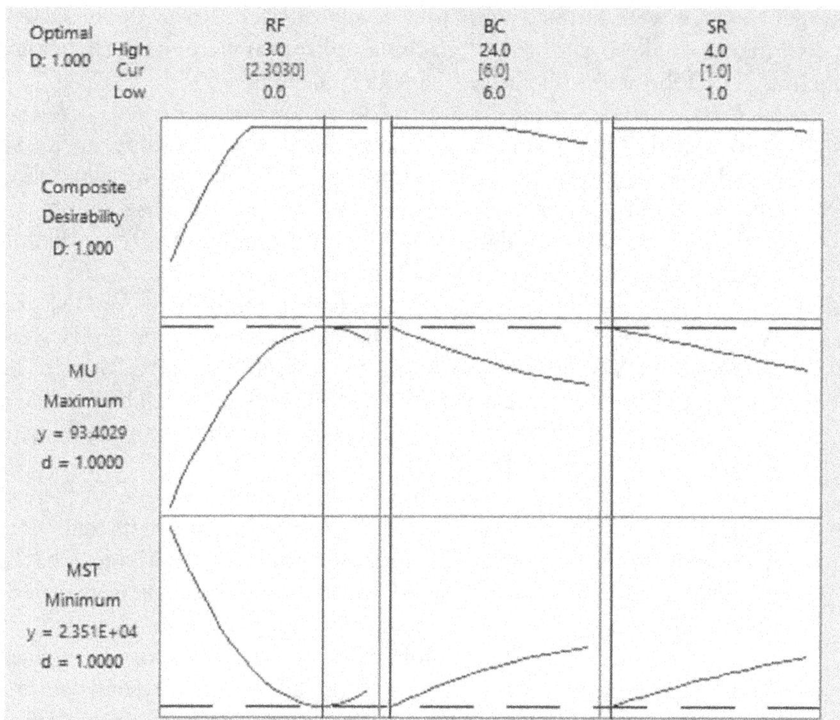

Figure 11.6 Multi-objective Optimization Plot for Performance Measures.

Table 11.8 Predicted and Confirmatory Results.

	Results	
System factor	*Predicted*	*Confirmation*
RF	2.30	2
BC	6	6
SR	1	1
MU (%)	93.40	92.13
MST (min)	23510.00	23832.09

min, respectively, as shown in Table 11.8. The findings of the confirmatory experiment were found to be rather close to the predicted value of responses.

11.6 CONCLUSION

In the present study, a multi-agent FMS scheduling was examined along with the influence of important system factors. The system factors considered were: routing flexibility, buffer capacity and sequencing rule. The system factors were studied with four levels of each. To evaluate system performance, makespan time and machine utilization were the performance metrics. The following conclusions have been drawn:

1. Two scheduling approaches: conventional and multi-agent based scheduling approaches were compared. The multi-agent scheduling approach outperformed the conventional approach. The results for both MST and MU are significantly better for the multi-agent scheduling than the conventional scheduling.
2. Following Taguchi's methodology, statistical analysis of ANOM and ANOVA have been carried out. Optimal system factor levels were identified for MST and MU. The routing flexibility speed is found to be the most significant parameter for MST as well as MU. The predicted optimal values are in good agreement with the simulation experiment values.
3. The multi-objective optimization of MST and MU has been performed using the desirability approach. The optimization results will help in making decisions related to implementation of multi-agent based scheduling systems. The predicted and the actual (simulation experiment) values are in good agreement with each other.
4. In the present study, only a limited set of system factors have been studied. The present work can be extended to other system factors, i.e., sequencing flexibility, part-mix flexibility, composite dispatching rules, etc. The studied scheduling model can be extended to integrated scheduling models; with the incorporation of design and planning factors like loading/unloading strategies, AGV, etc.

REFERENCES

Ahmad, S., Z. A. Khan, M. Ali, and M. Asjad. 2021. Geometric and Harmonic means based priority dispatching rules for single machine scheduling problems. *International Journal of Production Management and Engineering* 9, no. 2: 93–102. https://polipapers.upv.es/index.php/IJPME/article/view/15217

Ali, M., and S. Wadhwa. 2010. The effect of routing flexibility on a flexible system of integrated manufacturing. *International Journal of Production Research* 48, no. 19: 5691–5709. https://www.tandfonline.com/doi/abs/10.1080/00207540903100044

Barbosa, J., P. Leitão, E. Adam, and D. Trentesaux. 2015. Dynamic self-organization in holonic multi-agent manufacturing systems: The ADACOR evolution. *Computers in Industry* 66: 99–111. https://www.sciencedirect.com/science/article/abs/pii/S0166361514001894

Borangiu, T., S. Răileanu, T. Berger, and D. Trentesaux. 2015. Switching mode control strategy in manufacturing execution systems. *International Journal of Production Research* 53, no. 7: 1950–1963. https://www.tandfonline.com/doi/abs/10.1080/00207543.2014.935825

Chan, F. T. S., R. Bhagwat, and S. Wadhwa. 2008. Comparative performance analysis of a flexible manufacturing system (FMS): A review-period-based control. *International Journal of Production Research* 46, no. 1: 1–24. https://www.tandfonline.com/doi/abs/10.1080/00207540500521188

Derringer, G., and R. Suich. 1980. Simultaneous optimization of several response variables. *Journal of Quality Technology* 12, no. 4: 214–219. https://www.tandfonline.com/doi/abs/10.1080/00224065.1980.11980968

Fragapane, G., D. Ivanov, M. Peron, F. Sgarbossa, and J. O. Strandhagen. 2022. Increasing flexibility and productivity in Industry 4.0 production networks with autonomous mobile robots and smart intralogistics. *Annals of Operations Research* 308, no. 1–2: 125–143. https://link.springer.com/article/10.1007/s10479-020-03526-7

Gao, R., L. Wang, R. Teti, D. Dornfeld, S. Kumara, M. Mori, and M. Helu. 2015. Cloud-enabled prognosis for manufacturing. *CIRP Annals* 64, no. 2: 749–772. https://www.sciencedirect.com/science/article/abs/pii/S000785061500150X

Groover, M. P. 2016. *Automation, Production Systems, and Computer-Integrated Manufacturing*. Pearson Education India. https://books.google.co.in/books?id=DREwDwAAQBAJ&lpg

Hussain, M. S., and M. Ali. 2019a. Distributed control of flexible manufacturing system: Control and performance perspectives. *International Journal of Engineering, Applied and Management Sciences Paradigm* 54, no. 2: 156–162. http://www.ijeam.com/Published%20Paper/Volume%2054/Issue%202/23.pdf

Hussain, M. S., and M. Ali. 2019b. A multi-agent based dynamic scheduling of flexible manufacturing systems. *Global Journal of Flexible Systems Management* 20: 267–290. https://link.springer.com/article/10.1007/s40171-019-00214-9

Kagermann, H. 2014. Change through digitization—value creation in the age of Industry 4.0. In *Management of Permanent Change*, pp. 23–45. Wiesbaden: Springer Fachmedien Wiesbaden. https://link.springer.com/chapter/10.1007/978-3-658-05014-6_2

Khan, W. U., and M. Ali. 2015. Effect of sequencing flexibility on the performance of flexibility enabled manufacturing system. *International Journal of Industrial and Systems Engineering* 21, no. 4: 474–498. https://www.inderscienceonline.com/doi/abs/10.1504/IJISE.2015.072731

Khan, W. U., and M. Ali. 2019. Effect of routing flexibility on the performance of manufacturing system. *International Journal of Production Management and Engineering* 7, no. 2: 133–144. https://riunet.upv.es/handle/10251/125553

Lee, J., B. Bagheri, and H. Kao. 2015. A cyber-physical systems architecture for industry 4.0-based manufacturing systems. *Manufacturing Letters* 3: 18–23. https://www.sciencedirect.com/science/article/abs/pii/S221384631400025X

Leitão, P., N. Rodrigues, J. Barbosa, C. Turrin, and A. Pagani. 2015. Intelligent products: The grace experience. *Control Engineering Practice* 42: 95–105. https://www.sciencedirect.com/science/article/abs/pii/S096706611500091X

Li, H., Y. Wu, D. Cao, and Y. Wang. 2021. Organizational mindfulness towards digital transformation as a prerequisite of information processing capability to achieve market agility. *Journal of Business Research* 122: 700–712. https://www.sciencedirect.com/science/article/abs/pii/S0148296319306241

Liu, Y., L. Wang, X. V. Wang, X. Xu, and L. Zhang. 2019. Scheduling in cloud manufacturing: State-of-the-art and research challenges. *International Journal of Production Research* 57, no. 15–16: 4854–4879. https://www.tandfonline.com/doi/full/10.1080/00207543.2018.1449978

Mashhood, H., and M. Ali. 2022. A simulation study to evaluate the performance of FMS using routing flexibility. *International Journal of Simulation and Process Modelling* 18, no. 3: 222–236. https://www.inderscienceonline.com/doi/abs/10.1504/IJSPM.2022.126894

Monostori, L. 2014. Cyber-physical production systems: Roots, expectations and R&D challenges. *Procedia Cirp* 17: 9–13. https://www.sciencedirect.com/science/article/pii/S2212827114003497

Nascimento, D. L. M., V. Alencastro, O. L. G. Quelhas, R. G. G. Caiado, J. A. Garza-Reyes, L. Rocha-Lona, and G. Tortorella. 2019. Exploring Industry 4.0 technologies to enable circular economy practices in a manufacturing context: A business model proposal. *Journal of Manufacturing Technology Management* 30, no. 3: 607–627. https://www.emerald.com/insight/content/doi/10.1108/JMTM-03-2018-0071

Pach, C., Thierry B., Thérèse B., and D. Trentesaux. 2014. ORCA-FMS: A dynamic architecture for the optimized and reactive control of flexible manufacturing scheduling. *Computers in Industry* 65, no. 4: 706–720. https://www.sciencedirect.com/science/article/abs/pii/S0166361514000396

Phadke, M. S. 1989. *Quality Engineering Using Robust Design*. Englewood Cliffs, NJ: PTR Prentice-Hall. Inc. https://dl.acm.org/doi/abs/10.5555/525794

Qin, J., Y. Liu, and R. Grosvenor. 2016. A categorical framework of manufacturing for industry 4.0 and beyond. *Procedia CIRP* 52: 173–178. https://www.sciencedirect.com/science/article/pii/S221282711630854X

Rey, G. Z., T. Bonte, V. Prabhu, and D. Trentesaux. 2014. Reducing myopic behaviour in FMS control: A semi-heterarchical simulation–optimization approach. *Simulation Modelling Practice and Theory* 46: 53–75. https://www.sciencedirect.com/science/article/abs/pii/S1569190X14000069

Rossit, D., and F. Tohmé. 2018. Scheduling research contributions to smart manufacturing. *Manufacturing Letters* 15: 111–114. https://www.sciencedirect.com/science/article/abs/pii/S2213846317300871

Tran, L. V., B. H. Huynh, and H. Akhtar. 2019. Ant colony optimization algorithm for maintenance, repair and overhaul scheduling optimization in the context of industrie 4.0. *Applied Sciences* 9, no. 22: 4815. https://www.mdpi.com/2076-3417/9/22/4815

Trappey, A. J. C., C. V. Trappey, U. H. Govindarajan, A. C. Chuang, and J. J. Sun. 2017. A review of essential standards and patent landscapes for the Internet of Things: A key enabler for Industry 4.0. *Advanced Engineering Informatics* 33: 208–229. https://www.sciencedirect.com/science/article/abs/pii/S1474034616301471

Wadhwa, S., and K. S. Rao. 2002. Framework for a flexibility maturity model. *Global Journal of Flexible Systems Management* 3, no. 2–3: 45–55. http://giftsociety.org/download/gift-journal/3-2&3.pdf

Wadhwa, S., A. Singholi, and M. Ali. 2010. Evaluating the effect of part-mix and routing flexibility on FMS performance. *Global Journal of Flexible Systems Management* 11: 17–23. https://link.springer.com/article/10.1007/BF03396591

Wang, L., X. V. Wang, L. Gao, and J. Váncza. 2014. A cloud-based approach for WEEE remanufacturing. *CIRP Annals* 63, no. 1: 409–412. https://www.sciencedirect.com/science/article/abs/pii/S0007850614001176

Wang, P., R. X. Gao, and Z. Fan. 2015. Cloud computing for cloud manufacturing: Benefits and limitations. *Journal of Manufacturing Science and Engineering* 137, no. 4: 040901. https://asmedigitalcollection.asme.org/manufacturingscience/article-abstract/137/4/040901/375172

Wang, X. V., and L. Wang. 2014. From cloud manufacturing to cloud remanufacturing: A cloud-based approach for WEEE recovery. *Manufacturing Letters* 2, no. 4: 91–95. https://www.sciencedirect.com/science/article/abs/pii/S2213846314000145

Xu, X. 2012. From cloud computing to cloud manufacturing. *Robotics and Computer-Integrated Manufacturing* 28, no. 1: 75–86. https://www.sciencedirect.com/science/article/abs/pii/S0736584511000949

Zbib, N., C. Pach, Y. Sallez, and D. Trentesaux. 2012. Heterarchical production control in manufacturing systems using the potential fields concept. *Journal of Intelligent Manufacturing* 23: 1649–1670. https://link.springer.com/article/10.1007/s10845-010-0467-3

Chapter 12

Role of spiritual personality for effective management in smart manufacturing

Mohammad Anas and Mohammad Asjad

12.1 INTRODUCTION: BACKGROUND AND DRIVING FORCES

Manufacturing is the process of converting the raw material into the finished product with the Manufacturing sector plays a very significant role in the development society in general and a nation in a particular. Further, they contribute to the social and economic growth of country by manufacturing the good quality product but at the minimum possible cost. Its importance may further enhance when it's started shifting from traditional to the smart one, in order to sustain their goals and objectives as well their manufactured products. The important feature of smart manufacturing may leads to concise all the activities into a single umbrella. In such cases, the role and responsibilities of the teams may further increased and that too governed by the management.

Managements' role plays an important role in efficiency and effectiveness of the firm in general and the workers in a particular. The management is vital in grooming and nourishing the personal well-being that may directly and/or indirectly affect the skilled, talent and intellect of human workforce. Management duties may extend beyond formal bonding. Outside of the official contact, management may accompany team members on field trips, vocational training, cultural programs, spiritual events, etc. These programs are considered as pillars for improved work culture as they bear the responsibility of educating and training the teams.

The importance of these events also increases during the pandemic period, when the personals feel alone and insecure, in such case, the spiritual events are fruitful and leading to effective and efficient system to mitigate the pandemic affect. Even though from those activities many personal (from top management to shop floor worker) developed their own spiritual personality by acquiring the spiritual features.

Spiritual features, in contrast, are concerned further by means of individual subjective experiences, occasionally shared with others (cf. Zinnbauer, 1997). Maslow (1976) similarly differentiated, "the subjective and naturalistic religious experiences and attitude" (spiritual) from institutional organized religions. Amini et al. (2014) conducted a study on the relationship

DOI: 10.1201/9781003495314-12

between religiosity and forgiveness and reported that religious people has the tendency to forgive others. Similar findings were also reported in previous researches (Konstam et al., 2003; Mullet et al., 2003; Neto, 2007; Tangney et al., 2005; Worthington & Wade, 1999). Anas et al. (2015) conducted a study on the relationship between religious commitment and spiritual personality among adults. Religious commitment was found to be positively correlated with spiritual personality.

Forgiveness as a disposition can be defined as the propensity to forgive (Mullet et al., 2005). Forgiveness as a relationship characteristic is similar to intimacy and trust (McCullough & Witvliet, 2002). DiBlasio (1998) defines "Forgiveness as a change in willpower to release the person from malevolent behaviour toward an offender." Forgiveness has always been helpful in building strong, stable, and good relationships.

Mullet et al. (2003) studied religious involvement with forgiving personality. Kidwell (2010) explored the relationship between religious commitment and forgiveness through quantitative and qualitative study. Amini et al. (2014) conducted a study to examine the relationship between religiosity and forgiveness among students and reported that religious people has the tendency to forgive others. Similar findings were also reported in previous researches (Konstam et al., 2003; Mullet et al., 2003; Neto, 2007; Tangney et al., 2005; Worthington & Wade, 1999). Worthington et al. (2000) reported that religious commitment has the significant importance on forgiveness in various world religions, it may help persons forgive more willingly; those people who has dedicated toward their religion are more likely to follow the beliefs and faiths of their religions and have a tendency to forgive others. Worthington et al. (1996) talk about that merely under particular circumstances do an individual's religious beliefs influence their forgiveness of others.

Anas et al. (2015) conducted a study on the relationship between religious commitment and spiritual personality among adults. Religious commitment was found to be positively correlated with spiritual personality. Johnstone et al. (2012) examined the relationship among spirituality, religiosity, personality, and health care of different faith groups.

From the literature, the dimensions of spirituality are finalized and summarized in Table 12.1:

Table 12.1 Spiritual Personality Indicators

1	Forgiveness (F)
2	Religious commitment (R)
3	Gratitude (G)
4	Spiritual Personality Inventory (S)

In order to study the effect of spiritual personality, an attempt is made to rank and prioritize their rank with the help of Multi-Criteria Decision-Making (MCDM) technique is employed for ranking the spiritual personality indicators and/or enablers in order to understand the importance and severity of them.

12.2 F-AHP METHODOLOGY AND ITS IMPLEMENTATION

Zadeh designed and developed the fuzzy theory in the year of 1965, which helps to convert the managerial decision-making process into the linguistic variables. Bellman and Zadeh (1970) studied the possibility for the integration of fuzzy theory with multi-criteria decision-making in order to address the accuracy while estimating the weights. AHP has several flaws, such as its failure to resolve the confusion and/or ambiguity associated with mapping expert judgment into a precise number (Deng, 1999). This is because the AHP technique requires subjectivity in paired comparisons, and experts may be unable to assign precise values for comparison judgments (Chan & Kumar, 2007). As a result, vagueness-type uncertainty may take precedence in this process. Allowing decision-makers to choose from interval ranges rather than a single numeric value expressly gives them more confidence. Furthermore, most mathematical and computational tools demand precise values. These models are unable to account for the qualitative and confusing aspects of the decision-making process.

The associated inconsistencies and uncertainties can be addressed using fuzzy set theory (Bellman & Zadeh, 1970), which can evaluate qualitative and incomplete information (Kulak et al., 2005). As a result, fuzzy AHP is a hybrid technique that combines analytical hierarchy process with fuzzy logic theory. Fuzzy AHP can help to make smart and efficient decisions that may contain uncertainty (Gani et al., 2021). The steps involved in implementing fuzzy AHP (Saaty, 2008) for weightage evaluation are as follows (Gani et al., 2021; Gupta et al., 2023; Parmar et al., 2023; James et al., 2023).

Step 1 (Procedure): Create a pairwise comparison matrix based on the expert's recommendations.

Assume that the evaluation system contains "N" elements. The pairwise comparison of elements "*I*" and "*j*" yields a square "NxN" matrix "M_1." Each member e_{ij} in the matrix denotes the relative importance of element "*I*" in relation to element "*j*." When $i = j$, $e_{ij} = 1$ and $e_{ji} = 1/v$ for "M_1." For pairwise comparison of system elements, Saaty's 1–9 scale (Saaty, 2008) is used, as indicated in Table 12.2.

Step 1 (Implementation): Formation of pairwise comparison matrix based on the expert's opinion.

The methodology of FAHP is implemented for the assessment of weightage for each of the spirituality personality indicators as identified in the previous section. Before reaching out to industry professionals, an in-depth literature review was conducted on spirituality indicators, and it was concluded that a lot of information exists related to spirituality as a whole, but a comprehensive guide was missing which lists down the indicators that may affect the performance of manufacturing sectors in India.

Table 12.2 Numeric Value Allocation for Pairwise Comparisons of Matrix Elements

Verbal judgment of relative importance between elements "I" and "j"	*Numerical rating*
Equal importance	1
Moderate importance of one element over other	3
Strong importance	5
Very strong importance	7
Extremely strong importance	9
	2, 4, 6 and 8 are intermediate values

Table 12.3 Experts and Details

Expert	*Designation*	*Experience (in years)*	*Specialization*
Exp-1	General manager	22	Production Engineer
Exp-2	Deputy General Manager	17	Operations Management
Exp-3	Manager	12	Human Resource
Exp-4	Assistant Manager	9	Environmental Engineer
Exp-5	Technician	7	Maintenance Engineer

Table 12.4 Experts Judgment

Spirituality Personality	*F*	*R*	*G*	*S*
F	1	1/2	5	7
R	2	1	5	5
G	1/5	1/5	1	3
S	1/7	1/5	1/3	1

In this study, the survey method is utilized wherein various industry professionals specific to manufacturing sectors were approached, working under different capacities. The experts helped in validation of the literature review and findings, and helped framing the questionnaire more specific to the Indian context. The experts (as mentioned in Table 12.3) were given the list of criteria and requested to develop a pairwise comparison matrix of the purchase criteria based on Saaty's 1–9 scale. The developed matrix is shown in Table 12.4.

Step 2 (Procedure): Create a fuzzy pairwise comparison matrix for expert assessments (Tables 12.5 and 12.6).

Table 12.5 Linguistic Variables, Numerical Crisp Values, and Fuzzy Number

Linguistic variable	*Numerical crisp values*	*Fuzzy number*		
		l	*m*	*U*
Equal importance	1	1	1	1
Moderate importance of one element over other	3	2	3	4
Strong importance	5	4	5	6
Very strong importance	7	6	7	8
Extremely strong importance	9	9	9	9
Intermediate values	2	1	2	3
	4	3	4	5
	6	5	6	7
	8	7	8	9

Table 12.6 Linguistic Variables, Numerical Crisp Values and Reciprocal Fuzzy Number (RFN)

Linguistic Variable	*Numerical Crisp values*	*RFN*		
		l	*m*	*u*
Equal importance	1	1	1	1
Moderate importance of one element over other	1/3	1/2	1/3	1/4
Strong importance	1/5	1/4	1/5	1/6
Very strong importance	1/7	1/6	1/7	1/8
Extremely strong importance	1/9	1/9	1/9	1/9
Intermediate values	1/2	1/1	1/2	1/3
	1/4	1/3	1/4	1/5
	1/6	1/5	1/6	1/7
	1/8	1/7	1/8	1/9

The fuzzy pairwise comparison matrix is shown in expression (1)

$$\widetilde{X} = \begin{bmatrix} \tilde{x}_{11} & \tilde{x}_{12} & \cdots & \tilde{x}_{1n} \\ 1/\tilde{x}_{12} & \tilde{x}_{22} & \cdots & \tilde{x}_{2n} \\ \vdots & \vdots & \ddots & \vdots \\ 1/\tilde{x}_{1n} & 1/\tilde{x}_{2n} & \cdots & \tilde{x}_{nn} \end{bmatrix} \quad (1)$$

Where, all the diagonal elements in the above matrix are 1 i.e. $\tilde{x}_{ij} = (1,1,1)$ for $i = j$

Step 2 (Implementation): Based on the expert opinion, the fuzzy pairwise comparison matrices is developed with the help of Tables 12.5 and 12.6 and given in Table 12.7.

Step 3 (Procedure): Using the fuzzy geometric mean method, estimate the weight.

Table 12.7 Fuzzy-Based Experts Judgment

Spirituality Personality	*F*	*R*	*G*	*S*
F	(1,1,1)	(1, 1/2, 1/3)	(4,5,6)	(6,7,8)
R	(1/4,1/5,1/6)	(1,1,1)	(4,5,6)	(4,5,6)
G	(1/4,1/5,1/6)	(1/4, 1/5, 1/6)	(1,1,1)	(2,3,4)
S	(1/6, 1/7, 1/8)	(1/4, 1/5, 1/6)	(1/2, 1/3, 1/4)	(1,1,1)

Table 12.8 Fuzzy Geometric Mean

Spirituality Personality	*Fuzzy Geometric Mean Value*		
F	2.21	2.04	2.00
R	2.00	2.65	3.22
G	0.59	0.58	0.57
S	0.38	0.31	0.27
$(\tilde{r}_1 \oplus \tilde{r}_2 \oplus \tilde{r}_3 \oplus \tilde{r}_4)$	5.18	5.58	6.06

Calculate the weights of criteria using the geometric mean approach established by Buckley (1985), by following the stages (Kordi, 2008), in which the weights in a pairwise comparison matrix of alternatives are derived by expression (2). The fuzzy weight is determined by expression (3).

$$r_i = \prod_{j=1}^{n} (a_{ij})^{1/n} \tag{2}$$

at which a_{ij}, where (i,j=1, . . . n) are the comparison ratios in the pairwise matrix and "n" is the number of alternatives.

$$w_i = \frac{r_i}{\sum_j r_j} \tag{3}$$

Step 3 (Implementation): Calculate the weight using fuzzy geometric mean method

Fuzzy geometric mean is estimated based on the formula mentioned in Equation 2 and is mentioned in Table 12.8.

$$r_i = \prod_{j=1}^{n} (a_{ij})^{1/n} = (1\times1\times4\times6)^{\frac{1}{4}}; (1\times1/2\times5\times7)^{\frac{1}{4}}; (1\times1/3\times6\times8)^{\frac{1}{4}}$$

$$= (2.21,\ 2.05,\ 2)$$

Normalized weight, estimated as the center of area/total weight, is given in Table 12.10.

Step 4 (Procedure): Conduct a consistency check.

Table 12.9 Fuzzy Geometric Mean and Fuzzy Weight

Spirituality Personality	*Fuzzy Geometric Mean Value*	*Fuzzy Weight*
F	(2.21, 2.04, 2.00)	$(2.21, 2.04, 2.00) \otimes \left(\frac{1}{6.06}, \frac{1}{5.58.}, \frac{1}{5.18}\right)$
R	(2.00, 2.65, 3.22)	$(2.00, 2.65, 3.22) \otimes \left(\frac{1}{6.06}, \frac{1}{5.58.}, \frac{1}{5.18}\right)$
G	(0.59, 0.58, 0.57)	$(0.59, 0.58, 0.57) \otimes \left(\frac{1}{6.06}, \frac{1}{5.58.}, \frac{1}{5.18}\right)$
S	(0.38, 0.31, 0.27)	$(0.38, 0.31, 0.27) \otimes \left(\frac{1}{6.06}, \frac{1}{5.58.}, \frac{1}{5.18}\right)$

Table 12.10 Normalized Relative Weights of Purchase Criteria and Ranking

Spirituality Personality	*Fuzzy Weight*			*Center of Area (l+m+u)/3*	*Normalized weight*	*Rank*
F	0.364658	0.364881	0.385514	0.371684	0.36865	2
R	0.329505	0.474389	0.621392	0.475095	0.471217	1
G	0.097963	0.104999	0.111288	0.10475	0.103895	3
S	0.062592	0.055731	0.051783	0.056702	0.056239	4
Total weight				1.008231	1	

To ensure that expert judgments/opinions are consistent in nature, the consistency of each fuzzy pairwise comparison matrices must be checked. It is accomplished by de-fuzzing the fuzzy pairwise comparison matrix by assigning a crisp numerical value in place of the fuzzy numbers. The defuzzification in this work is performed through the use of the value graded mean integration approach (Khan et al., 2019; James et al., 2023).

Let a TFN is given by $T = (l,m,u)$, then the corresponding defuzzified crisp number is obtained by expression (4)

$$T_{crisp} = \frac{l+m+u}{3} \tag{4}$$

Followed by this, the pairwise matrix is normalized by dividing each element of a column by the sum of all the elements of the same column as given by expression (5).

$$\tilde{x}_{ij} = \frac{x_{ij}}{\sum_{j=1}^{n} x_{ij}} \qquad \forall i,j \qquad x_{ij} \in X \tag{5}$$

where, n is the number of criteria selected for the comparison.

The pairwise matrix is then normalized by dividing each element in a column by the total of all the elements of the same column, as provided by equation (5).

$$\tilde{x}_{ij} = \frac{x_{ij}}{\sum_{j=1}^{n} x_{ij}} \qquad \forall i,j \qquad x_{ij} \in X \tag{5}$$

where n is the number of criteria selected for the comparison.

The average value of the preference matrix for a criterion is used to generate a preference vector.

The Eigenvectors w_i, which indicate the weight of each criterion, are evaluated and are shown below.

$$w_i = \frac{\sum_{j=1}^{n} x_{ij}}{n} \tag{6}$$

Here, $\sum_{j=1}^{n} x_{ij}$ represents the sum of all the elements in row i of the normalized pairwise matrix. The largest eigenvalue λ_{max} is evaluated by expression (7)

$$\lambda_{max} = \sum_{i=1}^{n} \left\{ \left(\sum_{j=1}^{n} x_{ij} \right) \times w_i \right\} \tag{7}$$

The Consistency Index (CI) is evaluated by using expression (8) and Consistency Ratio (CR) as given by equation (9). Table 12.11 presents the random index RI for up to 10 criteria and it is used for the calculation of CR.

$$CI = \frac{\lambda_{max-} n}{n-1} \tag{8}$$

$$CR = \frac{CI}{RI} \tag{9}$$

When the CR is 0.10, it shows that the matrix's consistency is acceptable; otherwise, the pairwise comparison must be repeated until an acceptable CR of 0.10 is achieved.

Step 4 (Implementation): Evaluate the Consistency Index (CI)

Table 12.11 Random Index Values

Matrix	1	2	3	4	5	6	7	8	9	10
RI	0	0	0.58	0.90	1.12	1.24	1.32	1.41	1.45	1.49

Table 12.12 Preference Matrix

Spirituality Personality	*F*	*R*	*G*	*S*	*Preference Vector*
F	0.30	0.26	0.44	0.44	0.36
R	0.60	0.53	0.44	0.31	0.47
G	0.06	0.11	0.09	0.19	0.11
S	0.04	0.11	0.03	0.06	0.06

Table 12.13 Preference Vector and Corresponding Weight

Spirituality Personality	*F*	*R*	*G*	*S*	*Preference Vector*	*Corresponding weight*	*Value = (Corresponding weight/preference vector)*
F	1	1/2	5	7	0.36	1.56	4.35
R	2	1	5	5	0.47	2.04	4.35
G	1/5	1/5	1	3	0.11	0.46	4.14
S	1/7	1/5	1/3	1	0.06	0.24	4.04
Average							4.22
CI= (Avg-N)/(N-1)							0.072

The consistency index is evaluated as CI= (λmax—N)/(N-1). A lower value of the CI denotes meager consistency deviation.

Based on the initial Matrix, the Preference Matrix is developed as given in Table 12.12.

The next stage is to multiply the Pairwise comparison matrix with preference vector to get Corresponding weight;

Step 5 (Procedure): Establish the Random Index (RI) and Calculate the consistency ratio (CR)

Step 5 (Implementation): Establish the random index (RI) for the system elements with reference to Table 12.11, The RI is 0.9. The consistency ratio as CR = CI/RI. is 0.08, which is less than 0.10 and hence deemed to be fair, which indicates a reasonable perspective of the decision-makers in allocating the relative importance/weightage of elements (Parmar et al., 2023).

12.3 DISCUSSION AND CONCLUSION

The present research identified four indictors to gauge the spiritual personality in the smart manufacturing through literature survey. The first part of research work deals with the implementation of FAHP methodology, which estimates the weightages of distinct indicators that may further utilized to rank them. From Table 12.10, it can be concluded that Religiosity is the most important factor, whereas, forgiveness is the next one factor so as to get the spiritual personality among the personal involved in smart manufacturing, especially across the India. The next stage of the research deals with estimation of the consistency ratio, and from the analysis, it is found that consistency ratio is 0.08, which is less and deemed to be fair, that indicates a reasonable perspective of the decision-makers in allocating the relative importance/weightage of elements. Manufacturing sector has been greatly influenced by the Spiritual personality that has been discovered through literature analysis. In the analysis of indicator to Spiritual personality, the adoption of a hybrid technique (FAHP) helped to identify the most significant indicator. The FAHP method has assigned a weightage to each of the indicators. This method of prioritization of barriers would help vehicle owners, policy makers, the government, and potential entrepreneurs in the smart business to understand the importance of spiritual personality. The results of FAHP revealed that indicators, Religiosity, Forgiveness, Gratitude and Spiritual Personality Inventory are ranked from first to fourth. The suggested research and findings will serve as an instrument for the development and identification of spiritual personality, in smart manufacturing in a particular and other sector in a general. The study can be further extended to examine the inter-relationship between them through identification of sub factors of the said indicators with the help of different methodologies, say DEMATEL and Graph Theory, etc.

REFERENCES

Amini, F., Doodman, P., Edalati, A., Abbasi, Z. and Redzuan, M. 2014. A study on the relationship between religiosity and forgiveness among students. *Applied Science Reports*, 5(3):131–134.

Anas, M., Husain, A., and Aijaz, A. 2015. Relationship between religious commitment and spiritual personality among adults. *ACADEMICIA: An International Multidisciplinary Research Journal*, 5(11):181–189.

Bellman, R. E. and Zadeh, L. A. 1970. Decision-making in a fuzzy environment. *Management Science*, 17(4):B-141.

Buckley, J.J. 1985. Fuzzy hierarchical analysis. *Fuzzy Sets and Systems*, 17(3):233–247.

Chan, F.T. and Kumar, N. 2007. Global supplier development considering risk factors using fuzzy extended AHP-based approach. *Omega*, 35(4):417–431.

Deng, H. 1999. Multicriteria analysis with fuzzy pairwise comparison. *International Journal of Approximate Reasoning*, 21(3):215–231.

DiBlasio, F. A. 1998. The use of decision-based forgiveness intervention within intergenerational family therapy. *Journal of Family Therapy*, 20:77–94.

Gani, A., Asjad, M. and Talib, F. 2021. Prioritization and ranking of indicators of sustainable manufacturing in Indian MSMEs using fuzzy AHP approach. *Materials Today: Proceedings*, 46:6631–6637.

Gupta, S., Khanna, R., Kohli, P., Agnihotri, S., Soni, U. and Asjad, M. 2023. Risk evaluation of electric vehicle charging infrastructure using Fuzzy AHP–a case study in India. *Operations Management Research*, 16(1):245–258.

James, A. T., Asjad, M., Kumar, G., Shukla, V. C. and Arya, V. 2023. Analyzing barriers for implementing new vehicle scrap policy in India. *Transportation Research Part D: Transport and Environment*, 114:103568.

Johnstone, B., Dong Pil Yoon., Cohen, D., Schopp, L. H., McCormack, G., Campbell, J. and Smith, M. 2012. Relationships among spirituality, religious practices, personality factors, and health for five different faith traditions. *Journal of Religion Health*, 51:1017–1041.

Khan, A. A., Shameem, M., Kumar, R. R., Hussain, S. and Yan, X. 2019. Fuzzy AHP based prioritization and taxonomy of software process improvement success factors in global software development. *Applied Soft Computing*, 83:105648.

Kidwell, J. E. M. 2010. *Exploring the relationship between religious commitment and forgiveness through quantitative and qualitative study*. Ames, IA: Iowa State University.

Konstam, V., Holmes, W. and Levine, B. 2003. Empathy, selfism, and coping as elements of the psychology of forgiveness: A preliminary study. *Counseling and Values*, 47:172–183.

Kordi, M. 2008. *Comparison of fuzzy and crisp analytic hierarchy process (AHP) methods for spatial multicriteria decision analysis in GIS*. Master thesis. Gavle: University of Gavle.

Kulak, O., Durmuşoğlu, M.B. and Kahraman, C. 2005. Fuzzy multi-attribute equipment selection based on information axiom. *Journal of Materials Processing Technology*, 169(3):337–345.

Maslow, A. 1976. *Religions, values, and peak experiences*. New York: Penguin.

McCullough, M. E. and Witvliet, C. V. 2002. The psychology of forgiveness. In C. R. Synder and S. J. Lopez (ed). *Handbook of positive psychology* (pp. 446–458). New York: Oxford University Press.

Mullet, E., Barros, J., Frongia, L., Usai, V., Neto, F. and Shafigi, S. R. 2003. Religious involvement and the forgiving personality. *Journal of Personality*, 71:1–19.

Mullet, E., Neto, F., & Riviere, S. 2005. Personality and its effects on resentment, revenge, forgiveness, and self-forgiveness. In E. L. Worthington, Jr. (ed). *Handbook of forgiveness* (pp. 159–181). New York: Routledge.

Neto, F. 2007. Forgiveness, personality and gratitude. *Personality and Individual Differences*, 43(8):2313–2323.

Parmar, N. J., James, A. T. and Asjad, M. 2023. Analysis of maintenance outsourcing challenges for belt conveyors in industry 4.0 era. *Journal of Global Operations and Strategic Sourcing*, 16(3):718–744.

Saaty, T. L. 2008. Decision making with the analytic hierarchy process. *International Journal of Services Sciences*, 1:83–98.

Tangney, J. P., Boone, A. L. and Dearing, R. 2005. Forgiving the self: Conceptual issues and empirical findings. In E. L. Worthington, (ed). *Handbook of forgiveness* (pp. 143–158). New York: Routledge.

Worthington E. L. Jr. and Wade, N. G. 1999. The social psychology of unforgiveness and forgiveness and implications for clinical practice. *Journal of Social and Clinical Psychology*, 18:385–418.

Worthington, E. L., Jr., Kurusu, T. A., McCullough, M. E., & Sandage, S. J. 1996. Empirical research on religion in counseling: A ten-year review and research prospectus. *Psychological Bulletin*, 119:448–487.

Worthington, E. L., Sandage, S. J., & Berry, J. W. 2000. Group interventions to promote forgiveness: What researchers and clinicians ought to know. In M. E. McCullough, K. I. Pargament, & C. E. Thoresen (Eds.), *Forgiveness: Theory, research, and practice* (pp. 228–253). New York, NY: Guilford Press.

Zinnbauer, B. J. 1997. *Capturing the meanings of religiousness and spirituality: One way down from a definitional Tower of Babel*. Unpublished doctoral dissertation. Bowling Green State University.

Chapter 13

Smart manufacturing through lean enablers

An Assessment Framework Using Multi-Grade Fuzzy Approach and Importance Performance Analysis for Manufacturing Firm

Mohd. Aftab Alam, A. Abdullah, and Faisal Talib

13.1 INTRODUCTION

Lean manufacturing is a methodology that focuses on minimizing waste within manufacturing systems while simultaneously maximizing productivity (Parashar and Parashar, 2012). Lean principles share two basic standards: waste minimization and enhanced value. The purpose of this research is to assess lean enablers using a multi-grade fuzzy and importance performance analysis (IPA) method. Three enablers, 11 criteria, and 35 attributes were identified based on an exhaustive literature analysis and expert opinion. The multi-grade fuzzy approach is utilized in this study to measure the leanness index in the manufacturing industry. The importance performance analysis (IPA) technique is utilized to identify the weaker and stronger attributes of lean manufacturing in the organization that hinder the lean manufacturing overall performance. Theoretically, this study contributes by providing a paradigm for assessing leanness. This research can help manufacturing organizations discover essential lean practices on a regular basis and concentrate on their weaker attributes in order to advance their performance.

13.2 LITERATURE REVIEW

Lean manufacturing has become a popular approach for organizations seeking to improve their operations, reduce waste, and increase efficiency. The identification and implementation of lean enablers, which are factors that facilitate the adoption and implementation of lean principles, are critical to the success of lean manufacturing. The assessment of lean enablers is essential to identify areas for improvement and to ensure that organizations are fully leveraging the benefits of lean manufacturing. Therefore, a three-level (enablers, criteria, and attributes) model was developed to evaluate the leanness of the considered manufacturing firm. For the construction of the model, initially 3 enablers, 11 criteria, and 35 attributes were identified

DOI: 10.1201/9781003495314-13

through extensive literature reviews and expert opinions, as shown in Table 13.1.

This research chapter aims to assess lean enablers in a manufacturing firm using a multi-grade fuzzy approach and IPA. Pislaru et al. (2019) explored the use of fuzzy logic in measuring a company's performance. The research methodologies are based on principal component analysis, which employs language variables and criteria to produce quantitative measurements of long-term organizational success (Pislaru et al., 2019). The multi-grade fuzzy technique is used to avoid variable value instability and make computations easier (Ganesh and Suresh, 2016). Ganesh and Suresh (2015) implemented a

Table 13.1 Conceptual Model for Evaluating Leanness of a Firm

S. No.	*Category-I (Enablers)*	*Category-II (Criteria)*	*Category-III (Attributes)*
1.	Social and Environmental (O1)	Health and Safety (O11)	Indoor environmental health (O111)
			Occupational safety (O112)
			Workplaces standardization (O113)
		Employee Engagement (O12)	Defining roles of employees (0121)
			Empowering employees (O122)
			Incentives (O123)
		Environment Management (O13)	Eco-friendly practices (O131)
			Environmental policy (O132)
			Technological awareness (O133)
		Waste and Pollution management (O14)	Reduce, reuse, and recycle (O141)
			Air contamination (O142)
			Water contamination (O143)
			Land contamination (O144)
			Noise contamination (O145)
		Green Building (O15)	Site planning and design (O151)
			Energy efficiency (O152)
			Material selection (O153)
2.	Quality and Customer (O2)	Quality of products and process Quality (O21)	Members continually look for opportunities to improve the feature of the product/work (O211)
			Quality issues are captured and translated into training needs (O212)
			Quality related training requirements (O213)

Table 13.1 (Continued) Conceptual Model for Evaluating Leanness of a Firm

S. No.	*Category-I (Enablers)*	*Category-II (Criteria)*	*Category-III (Attributes)*
		Customer orientation (O22)	Both formal and informal feedbacks at all levels are taken from the customer (O221)
			Customer feedback/ suggestions are incorporated as early as possible (O222)
			Any change request from the customer is fully supported (O223)
3.	Technology (O3)	Manufacturing set-ups (O31)	Flexible manufacturing set-ups (O311)
			Less time for changing the machine set-ups (O312)
			Usage of automated tools used to enhance the production (O313)
		Product service (O32)	Products designed for easy serviceability (O321)
			Service centers well equipped with spares (O322)
			Minimum time required to restore the defective product to its original performance (O323)
		Design improvement (O33)	Management's interest toward evolving new models (O331)
			Training of design personnel in all aspects of design (O332)
			Preparedness of the management to invest on latest design techniques like RP and CAD/CAM (O333)
		Production methodology (O34)	Fully automated inspection systems (O341)
			Management's interest toward investment on FMS concepts (O342)
			Application of lean manufacturing principles for waste elimination (O343)

"triangular fuzzy logic approach" for estimating the threshold of safety level in a production company. Almutairi et al. (2019) used multi-grade fuzzy logic to evaluate the lean index for healthcare. Vinodh and Kumar (2012) created a multi-grade fuzzy leanness assessment decision support system to reduce computation time and error. Vimal et al. (2015) recognized that the process viewpoint is highly important and hence proposed quantifying the process manufacturing index by developing an inclusive framework using a multi-grade fuzzy approach.

Ainin and Hisham (2008) reformed this measuring framework for analyzing end-user satisfaction. IPA has been widely used to assess the service quality position of organizations with a view to identifying possibilities for service enhancement and long-term resource-based tactics (Deng, 2008). It is a popular instrument for managing techniques as it is easy, intuitive, and requires little awareness of statistical methods (Taplin, 2012; Anand and Kodali, 2010).

The present study found that these enablers were significant contributors to the successful implementation of lean manufacturing. It utilized a multi-grade fuzzy approach and IPA to assess lean enablers in a manufacturing firm. The multi-grade fuzzy approach is a comprehensive method for analyzing complex data sets and identifying key factors that influence organizational performance. IPA is a statistical tool used to identify the relative importance and performance of different factors in a particular context.

In conclusion, the assessment of lean enablers is critical to the success of lean manufacturing in organizations. Several studies have identified critical lean enablers such as leadership commitment, employee involvement, continuous improvement, and supplier involvement. This chapter has utilized a multi-grade fuzzy approach and importance-performance analysis to assess lean enablers in a manufacturing firm. The findings of this research can help organizations to identify areas for improvement and ensure that they are fully leveraging the benefits of lean manufacturing.

13.3 RESEARCH GAP

Based on the literature review, the following are a few potential research gaps that could be addressed by future studies:

1. Limited application of multi-grade fuzzy approach: While the multi-grade fuzzy approach has been identified as a comprehensive method for analyzing complex data sets, there is a lack of research that applies this approach to the assessment of lean enablers. Further studies could explore the use of multi-grade fuzzy approach in different industries and settings to assess lean enablers.
2. Lack of comparative studies: The literature review has identified several studies that assessed the impact of lean enablers on the implementation

of lean principles. However, there is a lack of comparative studies that evaluate the effectiveness of different assessment methods in identifying critical lean enablers. Future research could compare different assessment methods to identify the most effective approach for assessing lean enablers in a manufacturing firm.

3. Insufficient focus on specific lean enablers: While the literature review has identified several critical lean enablers, there is a lack of research that focuses on specific enablers such as supplier involvement or information sharing. Future studies could focus on specific enablers to identify their impact on the successful implementation of lean principles.
4. Limited attention to the role of cultural factors: The literature review has identified several organizational factors that facilitate the adoption of lean principles, but there is a lack of attention to cultural factors such as organizational culture or national culture. Future studies could explore the impact of cultural factors on the assessment of lean enablers in a manufacturing firm.

Overall, there are several research gaps that could be addressed by future studies, and by addressing these gaps, researchers can provide valuable insights that can help organizations to improve their operations and fully leverage the benefits of lean manufacturing.

13.4 RESEARCH OBJECTIVE

The objectives crafted from identified research gaps are as follows:

1. To identify the attributes that influenced the lean manufacturing in the organization through literature review and expert opinion.
2. To develop the lean index by measuring the leanness using multi-grade fuzzy approach in a case industry.
3. To identify the stronger and weaker attributes of lean manufacturing through IPA approaches in a case industry.
4. To develop a framework for assessing lean manufacturing system in manufacturing system.

13.5 METHODOLOGY

A three-phase methodology is provided to measure a manufacturing organization's leanness. To finalize the leanness practices, the first phase entails identifying experts, doing literature research, and consulting with them. For the study, six specialists with more than three years of expertise have been identified. The study identified three enablers, eleven criteria and thirty-five attributes through literature review. This factor has been validated by the six experts of the manufacturing industry. The thirty-five attributes were grouped into enablers, and the enablers were grouped into the theme criteria on the basis of literature review. The second phase entails gathering information from these experts via a standardized questionnaire with a 10-point rating scale ranging from "very worst" to "excellent." Each attribute, enablers and criteria were rated with a 10-point rating scale. Ratings and importance scale for the lean-index development is shown in Table 13.2. This study collected the responses from the employees of the case industry. The case manufacturing organization's leanness was determined in the third phase utilizing the multi-grade fuzzy technique and the performance ratings provided by these respondents. Using Importance Performance Analysis (IPA), this step also includes assessing the manufacturing organization's strong and weak attributes.

13.5.1 Multi-grade fuzzy approach

The multi-grade fuzzy approach is utilized in this study to measure the lean-manufacturing in the case manufacturing organization. The fuzzy logic method can be applied to practically any process that involves uncertainty, and it can be particularly useful at the organizational level for evaluating the performance of a model or prototype. When data is combined with uncertainties, fuzzy logic is a well-established tool that is utilized in the decision-making process. Pislaru et al. (2019) explored using fuzzy logic to measure a company's manufacturing performance. The research methodologies are based on principal component analysis, which employs language variables and criteria to produce quantitative measurements of long-term

Table 13.2 Ratings and Importance Scale for the Leanness

	Performance Rating Scale		*Importance Rating Scale (Weightage)*	
	Maximum is best for attribute		*Maximum is best for attribute*	
S. No.	*Linguistic variable*	*Performance Rating*	*Linguistic variable*	*Importance Rating*
1.	Very Worst (VW)	1	Extremely Low (EL)	1
2.	Worst (W)	2	Low (L)	2
3.	Very Poor (VP)	3	Below Medium (BM)	3
4.	Poor (P)	4	Medium (M)	4
5.	Below Average (BA)	5	Above Medium (AM)	5
6.	Average (A)	6	Moderately High (MH)	6
7.	Above Average (AA)	7	High (H)	7
8.	Good (G)	8	Very High (VH)	8
9.	Very Good (VG)	9	Extremely High (EH)	9
10.	Excellent (E)	10	Extremely Very High (EVH)	10

organizational success (Pislaru et al., 2019). The multi-grade fuzzy technique is used to avoid variable value instability and make computations easier (Jain et al., 2004; Ganesh and Suresh, 2016). In the industrial and service sectors, the multi-grade fuzzy was used to assess the effectiveness of "agile," "lean," "achievement," "safety practice level," and "supply chain management" (Vinodh and Aravindraj, 2015; Sridharan and Suresh, 2016; Ganesh and Suresh, 2016; Vinodh and Chintha, 2011; Vimal et al., 2015; Almutairi et al., 2019). Vinodh (2011) explained how to overcome disadvantages caused by conventional techniques through this approach. The technique of using the three dimensions of manufacturing to determine an organization's manufacturing level was also explored. Govindan et al. (2015) addressed why they chose the fuzzy approach rather than AHP. Ganesh and Suresh (2015) implemented a "triangular fuzzy logic approach" for estimating the threshold of safety level in a production company. Almutairi et al. (2019) used multi-grade fuzzy logic to evaluate the lean index for healthcare. Vinodh and Kumar (2012) created a multi-grade fuzzy leanness assessment decision support system to reduce computation time and error. Vimal et al. (2015) recognized that the process viewpoint is highly important and hence proposed quantifying the process manufacturing index by developing an inclusive framework using a multi-grade fuzzy approach.

13.5.2 Importance performance analysis

IPA technique is utilized by the study to identify the weaker and stronger attributes of lean manufacturing practices in the case organization. This enables the management to identify and provide attention to the weaker attributes to further develop and enhance for the productivity. The study also addresses and suggests for the improvement if the weaker attributes. The IPA is a widely used technique that generates a two-dimensional significance and efficiency grid in which the levels of significance and performance of various characteristics are plotted against one another, and the matrix is split into four quadrants (Martilla and James, 1977; Shieh and Wu, 2009). IPA is commonly used due to its well-known technique for professionals to assess existing approaches, identify emergency improvements for service attributes, and develop a new market strategy (Hansen and Bush, 1999). IPA has been widely used to assess the service quality position of organizations with a view to identifying possibilities for service enhancement and long-term resource-based tactics (Deng, 2008).

13.6 CASE STUDY ILLUSTRATION

13.6.1 Case organization

The study aims to understand and identify the weaker and stronger traits that will boost leanness of the manufacturing industry. A manufacturing organization which is in Aligarh Uttar Pradesh, India was chosen as a case study. The methodology is useful in offering a model for identifying lean-manufacturing practices in an organization. This case study was conducted in an Indian manufacturing industry which is in Uttar Pradesh state of India (hereafter referred to as XYZ manufacturing industry). XYZ is a manufacturing industry was established in the year 1971 by a visionary entrepreneur. This manufacturing industry is working in the field of automotive part solutions in South Asia having more than 50 years of innovation, technology, manufacturing and market leadership. XYZ manufacturing industry has become the provider of high-quality reliable parts such as Ignition Switches, Fuel Tank Caps, Casting Components and other automotive parts solutions worldwide. It has ultramodern manufacturing plants in India, located in Aligarh, Aurangabad, Pantnagar and Pune. The industry engages over 4000 plus direct and indirect personnel specialized in designing, production, quality and other associated services. XYZ industries strive to fulfil the ambitions of its customers by delivering top-notch engineering and manufacturing services and solutions. The company and its dedicated staff are persistently working toward effectively transforming all the abstract desires of their customers into tangible realities.

13.6.2 Sampling design and data collection

To collect data for this research, scheduled interviews with a closed-ended questionnaire about lean manufacturing on a 10-point rating scale ranging from "very worst" to "excellent" are used. For the multi-grade fuzzy approach, the importance/weightage for enablers, criteria, and attributes has been gathered from six specialists from related areas in which three are industrial experts while three are from academic background, including one project engineer, one managing director, one senior quality engineer, and three academic professors from different location of Aligarh, Uttar Pradesh. The expert's demography profile is shown in Table 13.3.

Table 13.3 Demographic Profiles of Experts

S. No.	*Background*	*Designation*	*Age (years)*	*Gender*	*Work experience (years)*	*Demography*
1.	Industrial (E1)	Project Engineer	27	Male	5	Aligarh, India
2.	Industrial (E2)	Managing Director	64	Male	32	Aligarh, India
3.	Industrial (E6)	Sr. Quality Engineer	28	Male	7	Aligarh, India
4.	Academic (E3)	Professor	54	Male	29	Aligarh, India
5.	Academic (E4)	Professor	49	Male	26	Aligarh, India
6.	Academic (E5)	Professor	50	Female	25	Aligarh, India

For the multi-grade fuzzy approach, the rating data has been gathered from six experts from the case organization in Aligarh. The experts are selected based on their expertise in the field related to manufacturing. Six experts from diverse industry and academic backgrounds have been selected to assess the importance rating, while other six experts from the case manufacturing industry have been chosen to evaluate the performance rating. The number of experts was chosen based on the organization's structure, the interviewees' expertise, and their involvement in lean manufacturing or continuous improvement efforts. Furthermore, several researchers have found that five respondents is an appropriate number (Elnadi and Shehab, 2016; Vinodh and Prasanna, 2011). In some research, fewer than five people are recruited (Behrouzi and Wong, 2013). Thus, meeting the above criteria, there were six experts available on the manufacturing unit. Thus, the study involves receiving responses from six experts. The experts have provided data for the current performance rating of the attributes of the case organization.

13.6.3 Assessment of leanness using multi-grade fuzzy approach

The case organization's lean assessment indicator is denoted by the letter I. It is the result of the experts' total assessment stage of rates based on each driver (B) and overall weights (W). The equation for a lean-manufacturing index is given by: $I = W \times B$ (Vinodh et al., 2010; Vinodh, 2011)

Since each factor involves a degree of uncertainty, the assessment scale has been divided into five levels. I = (10, 8, 6, 4, 2). 8–10 represents "extremely leanness," 6–8 represents "leanness," 4–6 represents "Fair leanness," 2–4 represents "not leanness" and less than 2 represents "extremely not leanness" this leanness scale is shown in Table 13.4. Six experts from various manufacturing organizations assigned the weightage. Attribute ratings are captured by six experts from the case organization based on their project operations. The attributes ratings scale and weightage scale are shown in Table 13.2 as already mentioned above (Azadeh et al., 2008; Vasanthan and Suresh, 2022, Vinodh et al., 2012). Table 13.5 shows the assessment model weights (performance ratings) as well as the expert's importance ratings.

Table 13.4 Leanness Scale

S. No.	*Linguistic variable*	*Rating levels (I)*
1.	Extremely not leanness	Less than 2
2.	Not leanness	2–4
3.	Fair leanness	4–6
4.	Leanness	6–8
5.	Extremely leanness	8–10

Table 13.5 Normalized Ratings

Enablers	*E1*	*E2*	*E3*	*E4*	*E5*	*E6*	*E-avg*	*E-avg. Sum*	*Weight (W)*
O1	8	7	9	7	7	9	7.833333	25.166667	0.311258278
O2	10	8	8	9	8	10	8.833333		0.350993377
O3	8	10	8	8	8	9	8.5		0.337748344

I. Importance (Weightage) Rating

Criteria	*E1*	*E2*	*E3*	*E4*	*E5*	*E6*	*E-avg.*	*E-avg. Sum*	*Weight (Wi)*
O11	8	9	8	8	9	9	8.5	39.833333	0.213389121
O12	8	9	7	6	6	9	7.5		0.188284519
O13	8	6	8	9	7	9	7.833333		0.19665272
O14	8	7	8	8	9	8	8		0.20083682
O15	8	8	9	7	8	8	8		0.20083682
O21	9	9	9	9	8	8	8.666667	17.166667	0.504854369
O22	9	9	8	9	8	8	8.5		0.495145631
O31	8	9	8	9	8	9	8.5	33.833333	0.251231527
O32	9	9	8	9	7	9	8.5		0.251231527
O33	9	9	8	8	8	8	8.333333		0.246305419
O34	8	10	8	9	8	8	8.5		0.251231527

Attributes	*E1*	*E2*	*E3*	*E4*	*E5*	*E6*	*E-avg.*	*E-avg. Sum*	*Weight (Wij)*
O111	4	9	8	9	9	8	7.833333	23.166667	0.338129496
O112	3	10	8	8	10	8	7.833333		0.338129496
O113	4	10	7	9	7	8	7.5		0.323741007
O121	6	10	7	7	6	7	7.166667	20.333333	0.352459016
O122	4	8	8	6	7	8	6.833333		0.336065574
O123	1	7	8	7	7	8	6.333333		0.31147541
O131	3	8	8	9	8	8	7.333333	22.833333	0.321167883
O132	6	8	8	7	7	8	7.333333		0.321167883
O133	5	9	9	8	9	9	8.166667		0.357664234
O141	6	10	9	9	9	8	8.5	38	0.223684211
O142	5	7	8	8	10	8	7.666667		0.201754386
O143	4	7	8	9	10	7	7.5		0.197368421
O144	6	7	7	7	9	7	7.166667		0.188596491
O145	4	9	8	9	6	7	7.166667		0.188596491
O151	7	9	8	6	8	8	7.666667	23.333333	0.328571429
O152	6	9	9	7	8	8	7.833333		0.335714286
O153	7	7	8	8	8	9	7.833333		0.335714286
O211	8	8	7	7	7	8	7.5	24.166667	0.310344828
O212	9	9	8	9	8	8	8.5		0.351724138
O213	8	8	8	9	8	8	8.166667		0.337931034

(Continued)

Attributes	*E1*	*E2*	*E3*	*E4*	*E5*	*E6*	*E-avg.*	*E-avg. Sum*	*Weight (Wij)*
O221	10	8	7	9	8	8	8.333333	25.166667	0.331125828
O222	9	9	8	9	8	8	8.5		0.337748344
O223	10	9	8	9	6	8	8.333333		0.331125828
O311	8	10	9	9	7	9	8.666667	26	0.333333333
O312	8	10	9	8	8	8	8.5		0.326923077
O313	9	10	8	9	8	9	8.833333		0.33974359
O321	8	10	8	9	8	8	8.5	24.333333	0.349315068
O322	6	10	7	9	8	8	8		0.328767123
O323	7	10	7	8	7	8	7.833333		0.321917808
O331	9	9	8	7	9	7	8.166667	23.666667	0.345070423
O332	8	8	7	8	7	8	7.666667		0.323943662
O333	4	10	9	9	7	8	7.833333		0.330985915
O341	2	6	9	9	7	7	6.666667	21.333333	0.3125
O342	1	9	9	9	7	8	7.166667		0.3359375
O343	3	9	8	8	9	8	7.5		0.3515625

II. Performance Rating

Oi	*Oij*	*Oijk*	*R1*	*R2*	*R3*	*R4*	*R5*	*R6*	*Wij*	*Wi*	*W*	*R-avg.*
O1	O11	O111	5	7	5	7	8	8	0.338129496	0.213389121	0.311258278	6.67
		O112	6	8	5	7	8	8	0.338129496			7
		O113	8	8	8	7	8	9	0.323741007			8
	O12	O121	8	8	8	8	7	9	0.352459016	0.188284519		8
		O122	7	5	7	6	6	4	0.336065574			5.83
		O123	4	4	4	5	5	4	0.31147541			4.33
	O13	O131	8	6	7	8	9	5	0.321167883	0.19665272		7.17
		O132	8	7	7	7	8	6	0.321167883			7.17
		O133	9	6	8	6	8	6	0.357664234			7.17
	O14	O141	9	7	9	8	9	6	0.223684211	0.20083682		8
		O142	8	6	9	7	9	8	0.201754386			7.83
		O143	7	8	9	7	9	6	0.197368421			7.67
		O144	8	8	9	6	9	8	0.188596491			8
		O145	8	8	8	7	8	7	0.188596491			7.67
	O15	O151	9	8	9	8	8	8	0.328571429	0.20083682		8.33
		O152	9	7	8	6	8	6	0.335714286			7.33
		O153	9	7	9	7	8	7	0.335714286			7.83
O2	O21	O211	9	7	8	6	7	7	0.310344828	0.504854369	0.350993377	7.33
		O212	8	8	7	7	8	8	0.351724138			7.67
		O213	9	7	7	7	8	8	0.337931034			7.67
	O22	O221	9	8	9	6	7	8	0.331125828	0.495145631		7.83
		O222	9	7	9	6	7	7	0.337748344			7.5
		O223	9	8	9	6	7	9	0.331125828			8

(Continued)

Table 13.5 (Continued) Normalized Ratings

Oi	*Oij*	*Oijk*	*R1*	*R2*	*R3*	*R4*	*R5*	*R6*	*Wij*	*Wi*	*W*	*R-avg.*
O3	O31	O311	8	6	9	5	6	8	0.333333333	0.251231527	0.337748 344	7
		O312	8	6	9	5	6	6	0.326923077			6.67
		O313	9	7	9	6	7	7	0.33974359			7.5
	O32	O321	9	7	10	6	9	7	0.349315068	0.251231527		8
		O322	8	7	8	7	7	6	0.328767123			7.17
		O323	8	5	8	6	7	5	0.321917808			6.5
	O33	O331	9	7	9	4	5	6	0.345070423	0.246305419		6.67
		O332	9	6	8	6	7	7	0.323943662			7.17
		O333	8	6	9	6	7	7	0.330985915			7.17
	O34	O341	8	5	8	5	6	6	0.3125	0.251231527		6.33
		O342	8	6	9	4	5	6	0.3359375			6.33
		O343	8	7	8	7	7	8	0.3515625			7.5

1. *Mean importance rating*: (*Sum of E-avg. of importance rating/35*) = 7.78
2. *Mean performance rating*: (*Sum of R-avg. of performance rating/35*) = 7.25

13.6.3.1 First-grade assessment calculation

The first-grade assessment of the "Health and Safety" (O11) criterion computation step is shown below:

The mean normalized weightage of six experts for the "Health and Safety" criterion is given as W_{11}= (0.338129496, 0.338129496, 0.323741007)

Assessment Vector for "Health and Safety" criterion is

$$B_{11} = \begin{bmatrix} 5\ 7\ 5\ 7\ 8\ 8 \\ 6\ 8\ 5\ 7\ 8\ 8 \\ 8\ 8\ 8\ 7\ 8\ 9 \end{bmatrix}$$

Index for "Health and Safety" criterion $I_{11} = \mathbf{W}_{11} \times \mathbf{B}_{11}$

I_{11} = (6.309352512, 7.661870496, 5.971223016, 6.999999993, 7.999999992, 8.323740999)

The index relating to other leanness requirements was derived using a similar process.

I_{12} = (6.418032786, 5.745901638, 6.418032786, 6.393442622, 6.040983606, 5.76229508)

I_{13} = (8.357664234, 6.321167883, 7.357664234, 6.963503649, 8.321167883, 5.678832117)

I_{14} = (8.02631579, 7.372807017, 8.811403509, 7.03508772, 8.811403509, 6.969298245)

I_{15} = (9.000000009, 7.328571436, 8.664285723, 6.99285715, 8.000000008, 6.99285715)

I_{21} = (8.648275862, 7.351724138, 7.310344828, 6.689655172, 7.689655172, 7.689655172)

I_{22} = (9, 7.662251656, 9, 6, 7, 7.993377484)

I_{31} = (8.33974359, 6.33974359, 9, 5.33974359, 6.33974359, 7.006410256)

I_{32} = (8.34931506, 6.356164377, 8.698630128, 6.328767117, 7.698630129, 6.027397254)

I_{33} = (8.669014085, 6.345070423, 8.676056338, 5.309859154, 6.309859154, 6.654929577)

I_{34} = (8, 6.0390625, 8.3359375, 5.3671875, 6.015625, 6.703125)

13.6.3.2 Second-grade assessment calculation

The second-grade assessment of the "Social and Environmental" enabler computation steps is shown below:

The mean normalized weightage of six experts for the "Social and Environmental" enabler is given as:

$$W_1 = (0.213389121, 0.188284519, 0.19665272, 0.20083682, 0.20083682)$$

Assessment Vector for "Social and Environmental" enabler is:

$$B_1 = \begin{bmatrix} 6.309352512, & 7.661870496, & 5.971223016, & 6.999999993, & 7.999999992, & 8.323740999 \\ 6.418032786, & 5.745901638, & 6.418032786, & 6.393442622, & 6.040983606, & 5.76229508 \\ 8.357664234, & 6.321167883, & 7.357664234, & 6.963503649, & 8.321167883, & 5.678832117 \\ 8.02631579, & 7.372807017, & 8.811403509, & 7.03508772, & 8.811403509, & 6.969298245 \\ 9.000000009, & 7.328571436, & 8.664285723, & 6.99285715, & 8.000000008, & 6.99285715 \end{bmatrix}$$

Index for "Social and Environmental" enabler is $I_1 = W_1 \times B_1$

I_1 = (7.617831929, 6.912477092, 7.439276784, 6.884229886, 7.857265779, 6.782019405)

Using a similar procedure, the index pertaining to other lean-manufacturing enablers has been derived:

I_2 = (8.822430532, 7.505480482, 8.146970204, 6.348175427, 7.348175427, 7.840041948)

I_3 = (8.337895051, 6.269640468, 8.677663872, 5.587751558, 6.592349223, 6.597684911)

13.6.3.3 Third-grade assessment calculation

The third-grade assessment of the "leanness index" computation step is shown below:

The mean normalized weightage of six experts for "leanness" is given as:

$$W = (0.311258278, 0.350993377, 0.337748344)$$

Assessment Vector for "leanness" is:

$$B = \begin{bmatrix} 7.617831929, & 6.912477092, & 7.439276784, \\ 8.822430532, & 7.505480482, & 8.146970204, \\ 8.337895051, & 6.269640468, & 8.677663872, \end{bmatrix}$$

$$\begin{matrix} 6.884229886, & 7.857265779, & 6.782019405 \\ 6.348175427, & 7.348175427, & 7.840041948 \\ 5.587751558, & 6.592349223, & 6.597684911 \end{matrix}\Bigg]$$

Leanness index $I = W \times B$

I = (8.28383818, 6.903500342, 8.105935668, 6.258194906, 7.251354957, 7.091119633)

$I = \frac{1}{6}$ (8.28383818 + 6.903500342 + 8.105935668 + 6.258194906 + 7.251354957 + 7.091119633)

I = 7.315657281 = 7.3 [(6–8) → "Leanness"]

The XYZ case organizations' leanness index is 7.3, which is indicates that XYZ is a "leanness" organization. Following that, the IPA was used to classify the attributes according to their importance and performance.

13.6.4 Classification of attributes using importance performance analysis

By applying IPA, the case organizations' lean-manufacturing attributes are classified based on their importance and performance. The diagrammatic representations of the four quadrants (Martilla and James, 1977; Tzeng and Chang, 2011) are shown in Figure 13.1. The horizontal axis is associated with attribute performance and the vertical axis is associated with attribute importance. IPA generally received backlash because of (a) the partition of the matrix and (b) the calculation of importance and performance (Abalo et al., 2007). Oh (2001) debated how crosshair positioning stays problematic, and those two options, scale-centric versus data-centric, are very popular. Taplin (2012) indicates that a large percentage of attributes lie in the quadrant of "keep up the good work" when observers use a scale-centered approach.

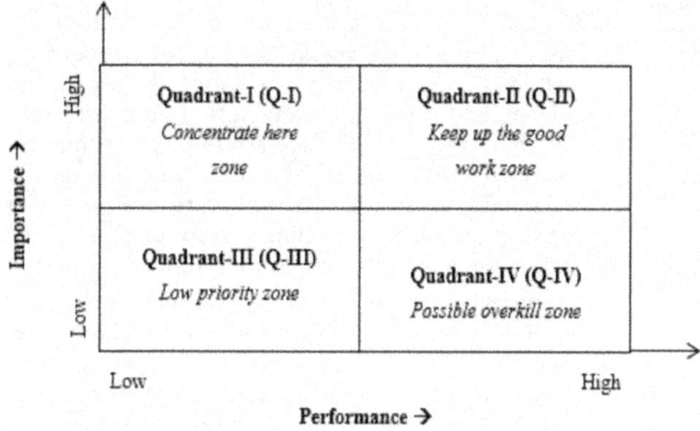

Figure 13.1 Diagrammatic Representation of IPA Quadrants.

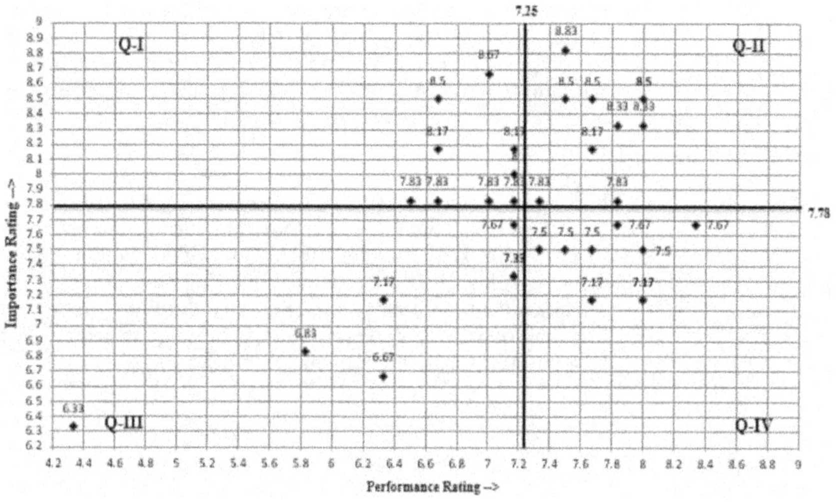

Figure 13.2 Analysis of Lean-Manufacturing Attributes.

13.6.5 IPA of the case organization

The x-axis in IPA represents attribute performance rating, whereas the y-axis represents importance rating. The case organizations' mean performance rating is 7.25, and the mean of importance rating is 7.78. Figure 13.2 indicates the case organizations' IPA of lean-manufacturing attributes. Furthermore, Table 13.6 provides a detailed explanation of each quadrant.

Table 13.6 Explanation of Quadrants in IPA

Quadrant-I	The quadrant holds significant importance, but its performance is lacking. Therefore, managers should take into account the quadrant's attributes in order to improve their performance. These attributes encompass indoor environmental health, occupational safety, eco-friendly practices, environmental policy, technological awareness, reduction, reuse, recycling, air contamination, water contamination, land contamination, and solvency.
Quadrant-II	The quadrant holds both high importance and high performance, so there is no need for any changes. It is crucial to maintain these good practices as they represent the firm's strengths. The attributes that contribute to this strength include realistic promise, client feedback, site planning, and design, material selection, visualization of the end project, skilled labour availability, communication and decision-making, and waste elimination.
Quadrant-III	The quadrant in question displays attributes that are of low importance and low performance. As a result, less emphasis should be placed on these factors. The attributes encompass workplaces standardization, incentives, noise contamination, predictability, flexibility, robustness, change management, pull inventory system, and lead time.
Quadrant-IV	The quadrant in question features attributes that are of low importance but high performance. Despite their lower importance, these attributes contribute significantly to the overall performance. The attributes include identification of desired outcomes, client focus, exceeding expectations, defining roles of employees, empowering employees, energy efficiency, profitability, liquidity, durability, service life, maintainability, process focus, project estimation, value stream mapping, and problem identification.

13.7 RESULTS AND DISCUSSION

The study highlights the importance of conducting a lean manufacturing evaluation to assess the current capacity of the case manufacturing organization. This evaluation is crucial in helping the organization enhance its lean manufacturing performance. The mean value of importance rating obtained from six different industrial and academic experts is 7.78, while the mean value of performance rating obtained from six employees of the case industry (XYZ) is 7.25, and these both values are used to draw all four quadrants of the IPA (performance rating versus importance rating). The study revealed that the organization achieved a value of 7.3, indicating its leanness status. However, it also identified nine weak attributes that require management's attention and improvement. All nine weak attributes are given below as per their decreasing criticality for the development of the lean manufacturing index of the case manufacturing industry in Table 13.7. The identified

Table 13.7 Suggestion for Improvement

S. No.	*Attributes (in decreasing critical attribute)*	*Explanation*
1.	Less time for changing the machine set ups (O312).	• Use of new technology and automation. • Standardize setup procedures. • Preparations, planning and parallel operations.
2.	Flexible manufacturing set-ups (O311).	• Design manufacturing layouts that allow for easy reconfiguration and adaptation to different production setups. • Continuous improvement and encourage employees to provide suggestions for enhancing the flexibility of manufacturing setups. • Train and cross-train your workforce to handle various manufacturing setups.
3.	Minimum time required to restore the defective product to its original performance (O323).	• Implement robust systems for early fault detection and diagnosis. • Invest in training and developing a skilled workforce with expertise in repairing the specific product or equipment. • Use of technology and automation to expedite the restoration process.
4.	Management's interest toward evolving new models (O331).	• Create a culture that encourages and values innovation within the organization. • Conduct workshops, training sessions, or seminars to educate management. • Establish mechanisms to recognize and reward individuals or teams.
5.	Indoor environmental health (O111).	• Giving employees light, temperature, and ventilation control. • Involve employees in discussion to find out whether they like the layout of the office.
6.	Technological awareness (O133).	• Conducting workshops and tie-ups with software companies for employees • Meeting and seminar
7.	Occupational safety (O112).	• Continuously train the workers in a standard manner. • Educate workers about important safety procedures. • Provide basic safety equipment.
8.	Service centers well equipped with spares (O322).	• Implement an efficient inventory management system to ensure accurate tracking of spare parts. • Use historical data and customer demand patterns to forecast future requirements for spare parts.

(*Continued*)

Table 13.7 (Continued) Suggestion for Improvement

S. No.	*Attributes (in decreasing critical attribute)*	*Explanation*
		• Implement a just-in-time delivery approach with suppliers for frequently used or critical spare parts. • Establish efficient repair and return processes to minimize the time required for repairing defective parts.
9.	Preparedness of the management to invest on latest design techniques like RP and CAD/CAM (O333).	• Communicate the benefits and advantages of adopting the latest design techniques. • Demonstrate the return on investment of implementing RP and CAD/CAM technologies • Offer training programs and workshops to familiarize management with the latest design techniques. • Long-term vision and roadmap for integrating RP and CAD/CAM technologies.

attributes encompass indoor environmental health (O111), occupational safety (O112), technological awareness (O133), flexible manufacturing set-ups (O311), less time for changing the machine set-ups (O312), service centers well equipped with spares (O322), minimum time required to restore the defective product to its original performance (O323), management's interest toward evolving new models (O331), and preparedness of the management to invest on latest design techniques like RP and CAD/CAM (O333). It's important to note that the current study focuses on evaluating a single manufacturing firm. However, the findings can be extrapolated and applied to other manufacturing firms as well. Moreover, both managers and researchers can utilize the framework established in this study to further enhance knowledge in the field of lean manufacturing. The framework serves as a foundation for future investigations and advancements. Table 13.7 provides specific recommendations for improving the weaker attributes identified in the study. This information can guide the efforts of managers and researchers in addressing the identified areas for improvement.

13.8 CONCLUSION

A total of nine weak attributes were identified among the 35 attributes, 11 criteria, and three enablers examined in this study. Lean manufacturing enables organizations to achieve a competitive edge, improve process flow and efficiency, enhance environmental sustainability, and enhance customer satisfaction by incorporating lean management principles and sustainable manufacturing practices (Ogunbiyi et al., 2014). It is a highly regarded and essential approach that all organizations must adopt to thrive in a competitive and unpredictable market. The objective of this study was to create a conceptual assessment framework for evaluating lean enablers in the manufacturing industry. The study examined the factors that impact lean manufacturing by employing two methodologies: the multi-grade fuzzy approach and importance performance analysis (IPA). The multi-grade fuzzy approach was used to evaluate the lean-manufacturing framework within the case organization. Weaker and stronger attributes were analyzed, assessed, and subsequently addressed using IPA. The findings revealed that the organization achieved a leanness status with a value of 7.3, indicating a solid understanding and implementation of lean manufacturing. However, there is room for improvement to enhance efficiency and productivity within the organization. The findings underscore the importance of evaluating the efficiency of manufacturing organizations through the lens of lean manufacturing. Such evaluation can aid the organization in enhancing its effectiveness in terms of lean principles. The project emphasizes that adopting the concept of lean manufacturing and implementing it can lead to improved performance within the organization, both internally and environmentally. Implementing lean practices within an enterprise requires time and is contingent upon multiple factors. Project managers and executive staff play a crucial role in fostering a positive mind-set, embracing organizational change, and formulating a plan for sustainable development. To test the generalizability of the study, the framework can be applied to other manufacturing firms situated in diverse regions.

REFERENCES

Abalo, J., J. Varela, and Manzano, V. 2007. Importance Values for Importance—Performance Analysis: A Formula for Spreading Out Values Derived from Preference Rankings. *Journal of Business Research* 60, no. 2: 115–121.

Ainin, S., and N. H. Haryati Hisham. 2008. Applying Importance-Performance Analysis to Information Systems: An Exploratory Case Study. *Journal of Information, Information Technology, & Organizations* 3: 95–103. https://doi.org/10.28945/132.

Almutairi, A. M., K. Salonitis, and A. Al-Ashaab. 2019. Assessing the Leanness of a Supply Chain Using Multi-Grade Fuzzy Logic: A Health-Care Case Study. *International Journal of Lean Six Sigma* 10, no. 1: 81–105. https://doi.org/10.1108/IJLSS-03-2018-0027.

Anand, G., and R. Kodali. 2010. A Mathematical Model for the Evaluation of Roles and Responsibilities of Human Resources in a Lean Manufacturing Environment. *International Journal of Human Resources Development & Management* 10, no. 1: 63–100. https://doi.org/10.1504/IJHRDM.2010.029447.

Azadeh, A., I. M. Fam, M. Khoshnoud, and M. Nikafrouz. 2008. Design and Implementation of a Fuzzy Expert System for Performance Assessment of an Integrated

Health, Safety, Environment (HSE) and Ergonomics System: The Case of a Gas Refinery. *Information Sciences* 178, no. 22: 4280–300. https://doi.org/10.1016/j.ins.2008.06.026.

Behrouzi, F., and K. Y. Wong. 2013. An Integrated Stochastic-Fuzzy Modeling Approach for Supply Chain Leanness Evaluation. *The International Journal of Advanced Manufacturing Technology* 68: 1677–1696.

Deng, W. J. 2008. Fuzzy Importance-Performance Analysis for Determining Critical Service Attributes. *International Journal of Service Industry Management* 19, no. 2: 252–70. https://doi.org/10.1108/09564230810869766.

Elnadi, M., and E. Shehab. 2016. A Multiple-Case Assessment of Product-Service System Leanness in UK Manufacturing Companies. *Proceedings of the Institution of Mechanical Engineers, Part B: Journal of Engineering Manufacture* 230, no. 3: 574–586.

Ganesh, J., and M. Suresh. 2015, Dec. Safety Practice Level Calculation in Indian Manufacturing Company Using Fuzzy Logic Approach. In *IEEE International Conference on Computational Intelligence and Computing Research (ICCIC)*, vol. 2015: 1–4. IEEE Publications. https://doi.org/10.1109/ICCIC.2015.7435777.

Ganesh, J., and M. Suresh. 2016, Dec. Safety Practice Level Assessment Using Multigrade Fuzzy Approach: A Case of Indian Manufacturing Company. In *IEEE International Conference on Computational Intelligence and Computing Research (ICCIC)*, vol. 2016: 1–5. IEEE Publications. https://doi.org/10.1109/ICCIC.2016.7919602.

Govindan, K., H. Soleimani, and D. Kannan. 2015. Reverse Logistics and Closed-Loop Supply Chain: A Comprehensive Review to Explore the Future. *European Journal of Operational Research* 240: 603–626. http://dx.doi.org/10.1016/j.ejor.2014.07.012.

Hansen, E., and R. J. Bush. 1999. Understanding Customer Quality Requirements: Model and Application. *Industrial Marketing Management* 28, no. 2: 119–130.

Jain, V., M. K. Tiwari, and F. T. S. Chan. 2004. Evaluation of the Supplier Performance Using an Evolutionary Fuzzy-Based Approach. *Journal of Manufacturing Technology Management* 15, no. 8: 735–44. https://doi.org/10.1108/17410380410565320.

Martilla, J. A., and J. C. James. 1977. Importance-Performance Analysis. *Journal of Marketing* 41, no. 1: 77–79. https://doi.org/10.2307/1250495

Ogunbiyi, O., J. S. Goulding, and A. Oladapo. 2014. An Empirical Study of the Impact of Lean Construction Techniques on Sustainable Construction in the UK. *Construction Innovation* 14, no. 1: 88–107.

Oh, H. 2001. Revisiting Importance-Performance Analysis. *Tourism Management* 22, no. 6: 617–627.

Parashar, A. K., and R. Parashar. 2012. Construction of an Eco-friendly Building Using Green Building Approach. *International Journal of Scientific & Engineering Research* 3, no. 6.

Pislaru, M., I. V. Herghiligiu, and I. B. Robu. 2019. Corporate Sustainable Performance Assessment Based on Fuzzy Logic. *Journal of Cleaner Production* 223: 998–1013. https://doi.org/10.1016/j.jclepro.2019.03.130.

Shieh, J. I., and H. H. Wu. 2009. Applying Importance-Performance Analysis to Compare the Changes of a Convenient Store. *Quality and Quantity* 43: 391–400.

Sridharan, V. and M. Suresh, M. 2016. Environmental Sustainability Assessment Using Multigrade Fuzzy—A Case of Two Indian Colleges. *IEEE International Conference on Computational Intelligence and Computing Research (ICCIC)*, IEEE Xplore 1–4. Chennai, India. https://doi.org/10.1109/ICCIC.2016.7919594.

Taplin, R. H. 2012. Competitive Importance-Performance Analysis of an Australian Wildlife Park. *Tourism Management* 33, no. 1: 29–37. https://doi.org/10.1016/j.tourman.2011.01.020.

Tzeng, G. H., and H. F. Chang. 2011. Applying Importance-Performance Analysis as a Service Quality Measure in Food Service Industry. *Journal of Technology Management & Innovation* 6, no. 3: 106–15. https://doi.org/10.4067/S0718-27242011000300008.

Vasanthan, P., and M. Suresh. 2022. Assessment of Organizational Agility in Response to Disruptive Innovation: A Case of an Engineering Services Firm. *International Journal of Organizational Analysis* 30, no. 6: 1465. https://doi.org/10.1108/IJOA-09-2020-2431.

Vimal, K. E. K., S. Vinodh, and R. Muralidharan. 2015. An Approach for Evaluation of Process Sustainability Using Multi-Grade Fuzzy Method. *International Journal of Sustainable Engineering* 8, no. 1: 40–54. https://doi.org/10.1080/19397038.2014.912254.

Vinodh, S. 2011. Assessment of Sustainability Using Multi-Grade Fuzzy Approach. *Clean Technologies & Environmental Policy* 13, no. 3: 509–15. https://doi.org/10.1007/s10098-010-0333-1.

Vinodh, S., and C. D. Kumar. 2012. Development of Computerized Decision Support System for Leanness Assessment Using Multi Grade Fuzzy Approach. *Journal of Manufacturing Technology Management* 23, no. 4: 503–16. https://doi.org/10.1108/17410381211230457.

Vinodh, S., and M. Prasanna, 2011. Evaluation of Agility in Supply Chains Using Multi-Grade Fuzzy Approach. *International Journal of Production Research* 49, no. 17: 5263–5276.

Vinodh, S., and S. Aravindraj. 2015. Benchmarking Agility Assessment Approaches: a Case Study [International journal]. *Benchmarking* 22, no. 1: 2–17. https://doi.org/10.1108/BIJ-04-2013-0037.

Vinodh, S., and S. K. Chintha. 2011. Leanness Assessment Using Multi-Grade Fuzzy Approach. *International Journal of Production Research* 49, no. 2: 431–45. https://doi.org/10.1080/00207540903471494.

Vinodh, S., S. R. Devadasan, B. Vasudeva Reddy, and K. Ravichand. 2010. Agility Index Measurement Using Multi-Grade Fuzzy Approach Integrated in a 20 Criteria Agile Model. *International Journal of Production Research* 48, no. 23: 7159–76. https://doi.org/10.1080/00207540903354419.

Vinodh, S., U. R. Madhyasta, and T. Praveen. 2012. Scoring and Multi-grade Fuzzy Assessment of Agility in an Indian Electric Automotive Car Manufacturing Organisation. *International Journal of Production Research* 50, no. 3: 647–60. https://doi.org/10.1080/00207543.2010.543179.

Appendix

1. QUESTIONNAIRE (IMPORTANCE RATING)

Objective: The purpose of this study is to propose a framework for assessing lean manufacturing in a firm: using a multi-grade fuzzy approach and importance-performance analysis methods. This study is being conducted by a student from the Department of Mechanical Engineering, Aligarh Muslim University.

Respondent's Profile

Name:
Age:
Gender:
Company/Department:
Position:
Experience (in years):
Qualification:

Weightage scale

S. No.	*Linguistic variable*	*Importance*
1.	Extremely Low (EL)	1
2.	Low (L)	2
3.	Below Medium (BM)	3
4.	Medium (M)	4
5.	Above Medium (AM)	5
6.	Moderately High (MH)	6
7.	High (H)	7
8.	Very High (VH)	8
9.	Extremely High (EH)	9
10.	Extremely Very High (EVH)	10

Note: On the basis of this scale table and using your experience and knowledge you are kindly requested to fill this form according to their importance, necessity and requirements of a manufacturing industry for assessing lean manufacturing. The scale ranges from Extremely Low (EL-1) to Extremely Very High (EVH-10).

Conceptual Model for Lean Enablers Assessment

S. No.	*Category-I (Enablers)*	*Importance*	*Category-II (Criteria)*	*Importance*	*Category-III (Attributes)*	*Importance*
1.	Social and Environmental		Health and Safety		Indoor environmental health	
					Occupational safety	
					Workplaces standardization	
			Employee Engagement		Defining roles of employees	
					Empowering employees	
					Incentives	
			Environment Management		Eco-friendly practices	
					Environmental Policy	
					Technological awareness	
			Waste and Pollution management		Reduce, Reuse and Recycle	
					Air contamination	
					Water contamination	
					Land contamination	
					Noise contamination	
			Green Building		Site planning and design	
					Energy efficiency	
					Material selection	
2.	Quality and Customer		Quality of products and process quality		Members continually look for opportunities to improve the feature of the product/work.	
					Quality issues are captured and translated into training needs.	
					Quality related training requirements.	

(*Continued*)

(Continued)

S. No.	*Category-I (Enablers)*	*Importance*	*Category-II (Criteria)*	*Importance*	*Category-III (Attributes)*	*Importance*
			Customer orientation		Both formal and informal feedbacks at all levels are taken from the customer.	
					Customer feedback/suggestions are incorporated as early as possible.	
					Any change request from the customer is fully supported.	
3.	Technology		Manufacturing set-ups		Flexible manufacturing set-ups.	
					Less time for changing the machine set-ups.	
					Usage of automated tools used to enhance the production.	
			Product service		Products designed for easy serviceability.	
					Service centers well equipped with spares.	
					Minimum time required to restore the defective product to its original performance.	
			Design improvement		Management's interest toward evolving new models.	
					Training of design personnel in all aspects of design.	
					Preparedness of the management to invest on latest design techniques like RP and CAD/CAM.	
			Production methodology		Fully automated inspection systems.	
					Management's interest toward investment on FMS concepts.	
					Application of lean manufacturing principles for waste elimination.	

2. QUESTIONNAIRE (PERFORMANCE RATING)

Objective: The purpose of this study is to propose a framework for assessing lean manufacturing in a firm: using a multi-grade fuzzy approach and importance-performance analysis methods. This study is being conducted by a student of the Department of Mechanical Engineering, Aligarh Muslim University.

Respondent's Profile

Name:
Age:
Gender:
Company/Department:
Position:
Experience (in years):
Qualification:

	Rating Scale	
	Maximum is best for attribute	
S. No.	*Linguistic variable*	*Performance Rating*
1.	Very Worst (VW)	1
2.	Worst (W)	2
3.	Very Poor (VP)	3
4.	Poor (P)	4
5.	Below Average (BA)	5
6.	Average (A)	6
7.	Above Average (AA)	7
8.	Good (G)	8
9.	Very Good (VG)	9
10.	Excellent (E)	10

Performance Rating Scale

Note: On the basis of this performance rating scale table and using your experience and knowledge you are kindly requested to fill this form based on performance, level of implication and practices of the given attributes in this industry. You have to just tick a right mark (✓) in the corresponding box according to their level of implication and performance of the attributes in your organization. The scale ranges from Very Worst (VW-1) to Excellent (E-10).

S. No.	*Category-I (Enablers)*	*Category-II (Attributes)*	*Performance Rating*									
			VW (1)	*W (2)*	*VP (3)*	*P (4)*	*BA (5)*	*A (6)*	*AA (7)*	*G (8)*	*VG (9)*	*E 10)*
1.	Social and Environmental	Indoor environmental health										
		Occupational safety										
		Workplaces standardization										
		Defining roles of employees										
		Empowering employees										
		Incentives										
		Eco-friendly practices										
		Environmental Policy										
		Technological awareness										
		Reduce, Reuse and Recycle										
		Air contamination										
		Water contamination										
		Land contamination										
		Noise contamination										
		Site planning and design										
		Energy efficiency										
		Material selection										
2.	Quality and Customer	Members continually look for opportunities to improve the feature of the product/work.										
		Quality issues are captured and translated into training needs.										
		Quality-related training requirements.										
		Both formal and informal feedbacks at all levels are taken from the customer.										

		Customer feedback/suggestions are incorporated as early as possible.
		Any change request from the customer is fully supported.
3.	Technology	Flexible manufacturing set-ups.
		Less time for changing the machine set-ups.
		Usage of automated tools used to enhance the production.
		Products designed for easy serviceability.
		Service centers well equipped with spares.
		Minimum time required to restore the defective product to its original performance.
		Management's interest toward evolving new models.
		Training of design personnel in all aspects of design.
		Preparedness of the management to invest on latest design techniques like RP and CAD/CAM.
		Fully automated inspection systems.
		Management's interest toward investment on FMS concepts.
		Application of lean manufacturing principles for waste elimination.

Index

L

M

N

O

P

Q

R

S

T

U

V

W

X

For Product Safety Concerns and Information please contact our EU representative GPSR@taylorandfrancis.com
Taylor & Francis Verlag GmbH, Kaufingerstraße 24, 80331 München, Germany

www.ingramcontent.com/pod-product-compliance
Lightning Source LLC
LaVergne TN
LVHW010855110826
845149LV00005B/1407

* 9 7 8 1 0 3 2 8 0 0 7 7 6 *